Aquarium Fishes

Other titles in the series

Wild Flowers
Mushrooms & Fungi
Birds
Minerals & Stones
Insects

Chatto Nature Guides

Aquarium Fishes

Illustrated and identified with colour photographs by

Siegfried Schmitz

Edited and translated by
Gwynne Vevers

Chatto & Windus London

Published by
Chatto & Windus Ltd.
40 William IV Street,
London WC2N 4DF

*

Clarke, Irwin & Co Ltd
Toronto

British Library Cataloguing in Publication Data
Schmitz, Siegfried
    Aquarium fish.—(Chatto nature guides).
    1. Aquarium fishes
    I. Title   II. Vevers, Gwynne
    639'.34   SF457

ISBN 0 7011 2355 9 Hardback
ISBN 0 7011 2356 7 Paperback

© BLV Verlagsgesellschaft—mbH, Munchen, 1977
English Translation © Chatto & Windus Ltd. 1978

Printed in Italy

Introduction

An aquarium is not just a tank of water with a few fishes swimming around. It is a man-made aquatic garden, in which plants and fishes are given conditions which are as close as possible to nature. The aquarist should strive to establish in his tank what is known as a biological equilibrium, so that both plants and animals so to speak collaborate. In a well-established aquarium, as in the wild, there should be a continuous cycle of events, which make any human interference almost unnecessary. Thus, the plants are not only there for decoration, for they also provide the fishes with some oxygen (at any rate during daylight hours). The fishes, on the other hand, produce carbon dioxide and faeces and these are used by the plants. In addition, some fishes use the plants as supplementary food, as shelter and as sites for spawning.

Choosing the right tank

The establishment of an aesthetically satisfying aquarium that is also in biological equilibrium requires a reasonably spacious tank, and this should be available from a good aquarium dealer. The moulded all-glass tanks sometimes seen are not really suitable, as they are very liable to break and the sides are not absolutely flat and so insufficiently transparent. Even less suitable are the spherical goldfish bowls still seen on the market. These should never be used.

The proper tank for home aquarium purposes is one with a frame of moulded angle-iron into which panes of suitably strong glass are fitted. The panes are held in place by a mastic and later, when the tank is filled, by the pressure of the water. Modern tanks of this type should have the angle-iron frame coated in nylon or some other suitable substance to prevent corrosion. Even more sophisticated are the elegant, frameless glass or plastic tanks in which the panes are glued directly to one another. A medium-sized tank is recommended for the beginner, and this could be about 70cm long, 30cm wide and 40cm tall.

A container of this size would be particularly suitable for use as what is known as a community tank, that is, one in which there are individuals of several different species. The present volume deals primarily with fish species which will thrive in a community tank because they are tolerant of one another and require more or less the same environmental conditions. The beginner should, however, be careful that in his initial enthusiasm he does not put too many fishes into a tank of this type. As a general rule many aquarists reckon 1cm of fish for each litre of water. A tank $70 \times 30 \times 40$cm contains approximately 80 litres. It could, therefore, house either 20 fishes each 4cm long or 40 fishes each 2cm long.

Angle-iron tank

All-glass tank

In contrast to the community tank is the species tank which is primarily used by specialists and breeders. This is intended for fishes of a single species, usually kept as a pair, which are not suitable for a community tank because they do not tolerate other species or because they require undisturbed surroundings for breeding. Small tanks are frequently suitable for this purpose.

Aquarium tanks can also be described in accordance with the types of fishes which they are intended to hold. All the species described in this book come from fresh waters, that is, from rivers, streams, ponds and inland lakes. There are two types of freshwater aquarium, namely the cold-water aquarium for fishes from temperate climates and the warm-water aquarium for all tropical freshwater fishes which must be kept in heated water. In addition there are marine or salt-water aquaria for those species which come from the sea, but these are not discussed in the present volume.

In summary, the beginner should gain his initial experience by using a medium-sized angle-iron or plastic tank, in which he will be able, over a period of years, successfully to maintain an attractive community of fishes and plants.

Positioning the tank

Tropical aquarium fishes require warmth and light if they are to thrive. The newly established tank could be on a window-sill where it would receive light (sometimes sunlight) the whole day. This would avoid the necessity for having artificial lighting and heating, but it would be to the detriment of the tank's occupants. Quite soon there would be a film of algae over the rocks and plants and on the tank glass, and the fishes would die because the sun would overheat the water.

It is, therefore, recommended that the tank should be fitted with its own lighting and heating so that it becomes completely independent of the sun. This also enables the tank to be positioned anywhere in the room, provided there is a nearby electric power point. An illuminated aquarium in a dark corner is also far more attractive than one in the vicinity of a window.

The tank can be placed on a strong table or on a low cupboard, but consideration should be given to its weight when filled with water. A medium-sized tank, as recommended for the beginner, will weigh about two hundredweight. In view of this it is better to use a purpose-built stand of timber or metal. For those who wish to hide the technical equipment there are on the market various aquarium units into which the tank and equipment can be fitted so that only the front glass is visible.

Cold-water tanks do not require heating and they can be positioned near a window, but still not directly on the window-sill. It is best to do this so that one of the ends of the tank faces an east or west window.

Setting up the tank

As soon as the tank is in position, remembering that it is not practical to move it, the aquarist can start setting it up. The first task is to put in the substrate. This should consist of coarse river sand or gravel (3-6mm) which can be bought from an aquarium dealer; under no circumstances should builders' sand be used. It must be carefully washed before it is put into the tank. This can be done by placing it into a plastic

Echinodorus

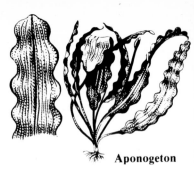

Aponogeton

Cryptocoryne

Cabomba

Elodea

Ludwigia

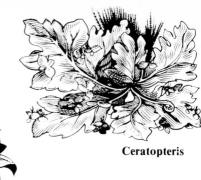

Ceratopteris

Sagittaria

Myriophyllum

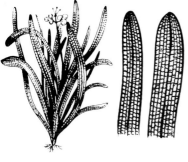

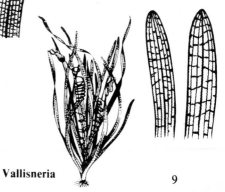

Vallisneria

bucket (in instalments) and running in water from the tap. While this is going on the gravel must be stirred so that the finer particles of dirt are washed away. The substrate does not need to contain any nutrients for the plants, but it provides a medium in which they can become rooted. The clean, wet gravel can now be placed in the tank as a layer 4-8cm thick so that its surface slopes gently towards the front of the tank. Small, decorative terraces can be built up using flat stones, but care should be taken not to introduce into the tank any rocks that contain metals or calcium, or those with sharp edges. Old, dead branches and roots which have lain for a long time in a river or lake can also be used as decorative material. These should be boiled in salt water and then rinsed in fresh water before being put into the tank.

When the substrate is in position the technical equipment can be installed in such a way that it does not spoil the appearance of the underwater landscape. The heating and lighting must not, of course, be switched on until the tank has been filled with water.

The plants can be put in when the tank is empty, but it is easier to do this if the tank has been filled to one-third of its depth with water. The most difficult method is to do this when the tank is full, as the plants will keep on floating to the top.

It is preferable to start with cheap, undemanding plants. After these have been purchased they must be kept submerged in a bowl of water until they are planted. At the start it is best to have only a few different kinds of plants, but to plant these in small groups. Large plants and those with long leaves should be planted at the back of the tank and smaller, delicate ones in the middle, leaving the foreground free as an area for swimming. Plants with large roots, which should be somewhat shortened, should be held between the index finger and thumb and inserted as deep as possible into the substrate. The gravel is then pressed down around them and the plants very gently pulled upwards so that the compressed roots become extended again. Unrooted cuttings are merely inserted in bunches into the gravel and anchored in position until they have rooted by glass rods, stones or pieces of lead (no other metals should be used).

Certain precautions should be taken when filling the tank with water, in order to avoid damaging the plants and stirring up the substrate. For an ordinary aquarium it is quite feasible to use tap water, but it should not be poured straight in as this would disturb the sand or gravel. It is better to introduce it by a length of tubing and allow it to splash over an upturned cup or over the back of the hand. Some aquarists spread a sheet of newspaper over the bottom before letting the water in; this can be easily removed when the tank is full.

The tank should be filled to within about 3cm of the upper edge. The plants can now be very carefully straightened. The water will probably be slightly cloudy and there may be small air bubbles on the glass but these will soon disappear.

The tank should be covered with a sheet of glass or with a cover supplied by the dealer. This restricts the rate of evaporation, prevents cooling of the water and also stops active fishes from jumping out. The tank should be left for at least a week, with the heater and thermostat switched on, before the first occupants are introduced. This period will allow the plants to start rooting, and the water to reach the correct temperature. In addition, and perhaps even more important, it will allow the mains water to lose any chlorine which would be dangerous to the fishes.

The newly purchased fishes must on no account be just tipped into the tank from the plastic bag in which they have been transported from the dealer. The bag should be allowed to float in the tank for about one hour so that any differences in temperature can be equalized. This should always be done when there is a risk of a sudden change of only a few degrees. At the end of about an hour the contents of the plastic can be slowly poured into the tank.

Technical accessories

The aquarist's technical equipment, available nowadays in a bewildering array, is intended solely to improve the living conditions of the fishes and plants.

A tropical aquarium tank will require lighting, a heater and thermometer, and an electric diaphragm pump connected with a filter and diffuser stone. A cold-water tank will only require the pump, filter and diffuser.

The tank should be lit with a special aquarium lamp which usually consists of a fluorescent tube with switch gear. Such fluorescent tubes are more expensive to buy than tungsten lamps, but they are cheaper to maintain, as they have a stronger life and use less current. Many of the aquarium tanks on sale have a cover with a special fitment for one or two fluorescent tubes. There are several different types of tube on the market and these not only give attractive colours but may also improve the growth of the fishes and plants. The lighting should normally be switched on for about 12 hours each day.

fluorescent tube

Nowadays aquarists only use an aquarium heater which can be regulated and controlled. The electric element, enclosed in a glass housing, is thermostatically controlled and has a switch knob for selecting the desired temperature. Once the correct temperature has been reached, it should stay constant for months, or even for years. In fact, the heating need never

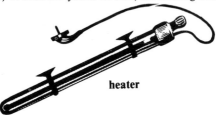

heater

be turned off. Nevertheless, it is advisable to have a thermometer attached by a suction cap to the inside of the glass panes. This will allow the aquarist to check the water temperature from time to time, so that he can if necessary make a slight adjustment by turning the switch knob. The heater itself

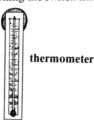

thermometer

should be attached vertically by two suction caps, preferably in an inconspicuous corner and in such a way that the rubber cap and switch knob are above the water level.

The diaphragm pump serves to move the water through the filter and also to release air through the diffuser stone. This latter function will release some oxygen into the water, but more importantly it will mix the different water layers, which might otherwise differ in temperature. The pump, which should run continuously, must be placed well above the tank so that in the event of an electric failure there is no risk of water being sucked back into it.

vibrator pump

There are two types of filter: internal and external. An internal filter is placed in an aquarium tank itself, whereas an external filter which is particularly suitable for a large tank, is usually a plastic container suspended outside the tank. In principle, both types work in the same way. The pump sucks the tank water through the filter, where the filter medium (nylon wool, activated charcoal, peat etc.) retains small particles of waste matter. It is, of course, necessary to renew the filter medium every week or month, depending upon the size and type of the filter. For a medium-sized tank it should be

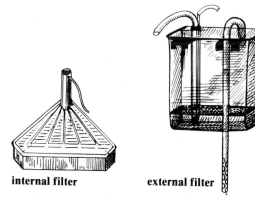

internal filter   **external filter**

sufficient to have an internal filter and a pump that can move 100-150 litres per hour.

By fitting a small plastic T-piece to the outlet tubing from the pump it will be possible to lead some of the air through a length of tubing to the porous diffuser stone. This can be placed in or on the substrate so that it releases a stream of small or large bubbles of air. The stream can be regulated by a small clamp fitted to the tubing between the T-piece and the diffuser stone.

diffuser stone

The whole of the technical equipment can be installed by the aquarist himself, provided he is reasonably practical and follows the instructions supplied by the manufacturers. If, however, he is in any doubt about his ability to do this he should bring in a professional who will be able to instal the equipment and possibly also advise on the choice of items available which is now bewildering, even to the experienced aquarist.

In addition to the basic technical equipment discussed above, there are certain other inexpensive items which are worth having:

1. A detritus sucker, basically a kind of pipette which can be attached to the suction of the pump and used, like a vacuum-cleaner, to remove coarse particles of detritus from the bottom of the tank.

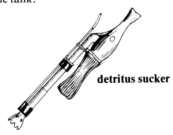

detritus sucker

2. A scraper which may be a razor blade or a piece of hard rubber mounted in a handle and used to remove algal films from the glass walls of the tank, particularly the front glass. To avoid scratches, a razor blade should not be used in a plastic tank.

scrapers

3. A plastic feeding ring which floats at the surface and serves to prevent the spread of dried food.

4. A food sieve, also of plastic, for feeding with live food, such as *Tubifex* worms. This prevents the worms from sinking to the bottom and creeping into the gravel where they die and cause pollution.

food sieve   **feeding ring**

Choosing the correct fishes

Since the middle of the last century, when the basic principles of aquarium maintenance were established in England, there has been a steady increase in the number of fish species found to be suitable for the aquarium. This progress has been greatly helped by advances in aquarium techniques and chemistry during recent years. Even though the home aquarist can only accommodate the smaller species, there are still hundreds of different fishes available to him. It is, therefore, not easy for the aquarist, particularly the beginner, to make a suitable selection. It is the aim of the present book to help the home aquarist select fish species which are not only attractive, hardy and not too expensive, but which also show interesting behaviour patterns. In addition to the wide range of tropical fishes there are also some species from temperate waters which can be kept in an aquarium.

The majority of aquarium fishes on the market have come, not from their original habitat, but from the tanks and ponds of specialist breeders. The relatively few wild-caught fishes imported are more suitable for the experienced aquarist as they are usually more delicate than captive-bred specimens, and also more expensive.

In choosing the first fishes for a community tank the aquarist must look for healthy specimens. They should exhibit normal behaviour, without rubbing their sides on the rocks or along the bottom, nor should they spend time at the surface gulping for air. Other points to look for are bright eyes, undamaged fins, a roundish body and a covering of scales without a film of white spots which would almost certainly be due to a dangerous disease.

Secondly, it is essential that the tank should not be over-crowded. Young or half-grown individuals purchased now will, in a few months' time, have doubled their length. It is best to start a new tank with a few young specimens of a few species.

It is a mistake to buy a single fish of one species. Sometimes it may be practical to buy a pair, but in general it is much better, particularly with shoaling fishes, to buy a small number of individuals of each species. It is usually advisable to buy fishes that are all of approximately the same size to avoid the risk of larger individuals eating the smaller ones.

Finally, some account should be taken of the general habits of the fishes to be purchased. Ideally the tank should have a mixture of bottom-living fishes, shoaling fishes that live in the middle water layers and surface-living species. In this way there would be proper utilization of the available space.

Scientific names

The beginner may find difficulty in understanding the scientific names he comes across in the literature and also when buying his fishes. Yet the scientific names of animals and of plants are very valuable and practical because they are completely international, and therefore understood throughout the world. The popular names, on the other hand, are not so widely known, and their use frequently leads to misunderstanding.

The practice of giving every animal and plant a double name, in Latin, dates back to the 18th century when the great Swedish naturalist Carl von Linné, usually known as Linnaeus, established his very logical binary nomenclature.

A species of animal or plant comprises all those individual

organisms which show the same characteristics and which breed with one another. Several closely related species are collected together or classified into a genus (plural genera), and several genera into a family. Families are then grouped into orders, orders into classes, and classes into phyla (singular phylum).

Thus the fish known colloquially as a Sumatra barb is classified as follows:

Phylum Chordata
  Class Osteichthyes (bony fishes)
    Order Cypriniformes
      Family Cyprinidae (barbs and carps)
        Genus *Barbus* (barbs)
          Species *Barbus tetrazona*, the Sumatra barb.

The "great families"

Within this system of zoological classification there are, in fact, dozens of families, but only a few of these are of importance to the aquarist. During the last hundred years it has been established that a number of fishes are suitable for the private aquarist, but many of these are unsuitable for the beginner because they are difficult to keep or require special conditions. Most of the fishes normally kept in a home aquarium belong to quite a small number of families, and the following remarks are intended to help the aquarist who is confronted with the task of stocking his first aquarium tank.

Apart from a few interesting species from other families, which will become known as the aquarist gains experience, the majority come from the following families:

1. **Characins** (Characidae) (pages 24-43). This is a large family with representatives in central and southern America and in tropical Africa. Most of the species have a characteristic adipose fin, which has no rays, and lies between the dorsal and caudal fins. Characins are attractively coloured, active and peaceful fishes which are best kept in small shoals. They require plenty of space for swimming and some shelter in the form of dense vegetation. They will all eat dried food, but a supplement of live food is much appreciated.

2. **Barbs and carps** (Cyprinidae) (pages 44-57, 134-139). A large family represented in Europe, North America, Africa and Asia. Most of the species are sociable, lively and peaceful. They are omnivorous and very suitable for the beginner because they do not require any special water conditions.

17

3. **Armoured catfishes** (Callichthyidae) (pages 58-67). These are small catfishes from South America. Although not particularly beautiful they are excellent for the community tank because they spend their time on the bottom picking up scraps of waste and thus helping to keep the tank clean. They rise to the surface from time to time to take in a small gulp of air, from which they are able to utilize the contained oxygen. They do this by means of an accessory method of respiration which supplements the work of the gills.

4. **Egg-laying toothcarps** (Cyprinodontidae) (pages 68-85). A family with members in southern Europe, Africa, Asia and America. Many are brightly coloured and very suitable for the aquarium but some should not be kept in a community tank, as they will eat smaller members of other species. The males fight one another and are rather aggressive towards the females. The tank should therefore have some dense vegetation to provide hiding-places. Each male can be kept with a small number of females. Most of the species live close to the water surface.

5. **Livebearing toothcarps** (Poeciliidae) (pages 86-93). This family is restricted to tropical and subtropical America. Many of the species have become very popular fishes and these include such household names as guppies, mollies and swordtails. Most are bred in large numbers, and are therefore not expensive, besides being peaceful and very adaptable. Whereas the females of the Cyprinodontidae lay eggs, those of the Poeciliidae produce live young.

6. **Cichlids** (Cichlidae) (pages 94-107). A family with numerous species in tropical America and Africa, but only a few in Asia. They are mostly attractive in coloration and pattern but their real interest to the aquarist lies in their breeding habits and territorial behaviour. They do, however, have some disadvantages. Although some, such as the angelfishes, can be kept in a community tank, the larger species are often quarrelsome and only to be housed as a pair in a species tank. They also have a propensity for digging up the bottom and eating the decorative plants. They are, in fact, omnivorous.

7. **Labyrinth fishes** (Anabantidae) (pages 108-117, 140-141). A relatively small family with representatives in tropical Asia and Africa. They have an accessory respiratory organ, the labyrinth, which enables them to make use of air taken in at the surface, thus allowing them to live in waters deficient in oxygen. They can literally drown if prevented from coming to

the surface to take in air. Most of the species are hardy, adaptable and very suitable for a community tank. They are easy to feed and tolerant of other species, although not always of members of their own species.

Feeding

Aquarium fishes should be fed a varied diet, but sparingly. Many beginners tend to feed too often and too richly. This results in obesity and the water becomes polluted with excess food. Many aquarium fishes must have died from overfeeding, but only a few from starvation. They should be fed two or three times a day and only given as much as they will consume in a few minutes.

There are two principal types of food: live and dried.

Live food, such as the small red worm Tubifex, water-fleas and mosquito or midge larvae can be bought from some aquarium dealers or it can be collected from ponds or bred by the aquarist himself. There is, however, a danger that food animals collected in the wild may introduce diseases into the aquarium. This does not happen if the live food has been deep-frozen or freeze-dried.

Dried food in the form of flakes, pellets or powder is now produced commercially by many firms and is of excellent quality. It contains all the important components of a balanced diet (fats, proteins, vitamins, mineral salts, trace elements). Many fishes, and indeed most of those described in this book, will remain healthy for years when fed exclusively on good dried food. Nevertheless they should from time to time be given a little live food.

Finally, some popular species, e.g. livebearers, labyrinth fishes and some barbs, require a supplement of greenstuff. This can be in the form of boiled lettuce and spinach or it can be purchased in dried form.

Health

Fish diseases are rather rare when the aquarium is well maintained, but the following points are important:

1. Not too many fishes, but plenty of plants.
2. Avoid overfeeding.
3. Check the temperature regularly and ensure that all the technical equipment is working properly.
4. Once a week remove any algal growths as well as detritus and plant fragments.

5. Make good losses due to evaporation by adding mains water at the correct temperature.

6. At intervals renew about one-fifth of the tank water.

In spite of these precautions there are times when disease strikes. Unfortunately when a fish is discovered to be ill it is usually too late to do anything about it. It should, however, be immediately isolated from the other occupants of the tank, otherwise they too may become infected. If the beginner cannot diagnose the trouble he should seek help from an experienced aquarist or a dealer. Any contacts with the local aquarist society should prove invaluable.

In some cases, however, the symptoms may be clear-cut and the following points are worth noting. General debility results in the fish remaining the whole time at the surface. This is due to oxygen lack owing to overcrowding or to polluted water (the results of insufficient aeration, decomposition of excess food, an overloaded filter, etc.). In such cases the deficiency can be put right and a proportion of the water renewed. Loss of appetite and twitching of the fins are also signs that all is not well, and they usually suggest the presence of parasites or an insufficiently high temperature. Parasites can sometimes be treated with drugs obtainable from a dealer, and the temperature can, of course, be adjusted.

Fungal diseases are not uncommon, particularly when a fish has suffered some external injury. These appear in the form of a white film on the body and fins. They sometimes disappear when the temperature is raised slightly and a proportion of the water is renewed. Alternatively the diseased fish can be caught and held in a net while the infected areas are carefully dabbed with diluted tincture of iodine or with some other preparation.

One of the most feared and infectious fish diseases is known as white-spot. This is caused by the parasite *Ichthyophthirius* and it appears in the form of tiny spots on the skin. At the first sign this condition must be treated immediately, using one of the special preparations that are available on the market.

Parasites of all kinds are indeed the worst enemies of aquarium fishes. Many of them are very difficult for the layman to identify and even more difficult to eradicate. One of the best ways of avoiding the introduction of parasites is to set up a small quarantine tank, complete with heating and aeration, in which all newly purchased fishes can be carefully

observed over a period of several days. If they show no ad-
verse signs they can then be introduced into the community
tank.

Aquarium Fishes

Red-nosed Tetra *Hemigrammus rhodostomus*

Characidae

Characteristics: up to 5cm. An attractive, remarkably slender fish with well-marked coloration. The caudal fin has three broad dark bands and the head is an intense red when the fish is in good condition. The sexes are difficult to distinguish, but the females are usually somewhat larger and stouter than the males. Like most characins these are shoaling fishes, which live mainly in the middle water layers. They are at first somewhat sensitive but usually become quite hardy once they are acclimatized. They are active but very tolerant of other fishes.—**Distribution:** South America, lower Amazon.—**Maintenance:** community tank from 30-40cm long, with space for swimming and some dense vegetation. This allows them to rest among the plants when they are not swimming. They must be kept as a shoal, as single individuals do not thrive. Water temperature about 25°C.—**Diet:** omnivorous with a preference for live food, particularly insects. A supplement of greenstuff is recommended.—**Breeding:** very difficult.

The genus *Hemigrammus* contains a number of popular aquarium fishes. Only two characteristic representatives are described here, but the other species are similar in behaviour and requirements. Of these the following are suitable for the beginner: Buenos Aires tetra *H. caudovittatus*, beacon fish *H. ocellifer* and pretty tetra *H. pulcher*.

Glowlight Tetra *Hemigrammus erythrozonus*

Characidae

Characteristics: up to 4.5cm. Easily recognized by the narrow, red fluorescent band running along each flank. Both sexes have the same coloration, but adult females are slightly larger and stouter. Peaceful, shoaling fishes of the lower water layers, where they frequently remain motionless among the plants.—Distribution: north-eastern South America, particularly Guyana.—**Maintenance:** as a shoal in a well planted community tank, with subdued lighting and clear water. Floating plants help to shade the light. Water temperature about 24°C.—**Diet:** dried food of all types with some greenstuff. Also small live food (*Tubifex*, water-fleas etc.) **Breeding:** not very easy.

Flame Tetra *Hyphessobrycon flammeus*

Characidae

Characteristics: up to c. 4cm. In the males the ventral and anal fins are a more intense red than in the females, and they have dark borders. Both sexes have two characteristic dark bars behind the gill covers.—**Distribution:** South America, around Rio de Janeiro.—**Maintenance:** always as a shoal in a community tank with numerous plants and open water for swimming. A particularly suitable, hardy fish for the beginner. Water temperature 20-25°C.—**Diet:** can be dried food only.—**Breeding:** possible even in hard water. Individuals ready to spawn (preferably a male and two or more females) should be put into a separate tank. The parents may try to eat their own eggs so they must be removed from the tank immediately after spawning.

Serpae Tetra *Hyphessobrycon serpae*

Characidae

Many authorities regard this as a subspecies of the very similar blood characin *Hyphessobrycon callistus* (see p. 30) and it is then called *H. callistus serpae*. The nomenclature of the blood characins is still in some confusion.— **Characteristics:** up to 4.5cm. The males are bright orange, the females somewhat paler. There is usually a small vertical band behind the gill cover. These are active but peaceful shoaling fishes which live mainly in the middle and lower water layers. —**Distribution:** central South America.—**Maintenance:** as a shoal in a community tank with dense vegetation and a large area of open water for swimming. The water should be kept at a temperature of about 24°C, but its composition is not critical.—**Diet:** omnivorous.—**Breeding:** not too difficult.

Griem's Tetra *Hyphessobrycon griemi*

Characidae

Characteristics: up to 4cm. Similar to the flame tetra but paler and rather more delicate. The sexes are difficult to distinguish but the males are usually smaller and more slender. These are peaceful shoaling fishes living in the middle water layers.—**Distribution:** central Brazil.—**Maintenance:** in a small community tank with plenty of open water for swimming and clumps of dense vegetation. Water temperature about 24°C, or a little lower.—**Diet:** all kinds of dried food and some small live food, with a supplement of lettuce. —**Breeding:** not difficult.

Other recommended species of *Hyphessobrycon* include the yellow tetra, *H. bifasciatus*, the flag tetra, *H. heterorhabdus* and the bleeding heart tetra, *H. rubrostigma*.

Lemon Tetra *Hyphessobrycon pulchripinnis*

Characidae

Characteristics: up to 5cm. An almost translucent yellowish fish. The sexes are difficult to distinguish but the female is somewhat larger with a higher back. The edge of the anal fin is blacker in the male. These are small, shoaling fishes which form territories among the plants.—**Distribution:** South America (Amazon).—**Maintenance:** as a shoal in a community tank with some dense vegetation to provide shelter, but not with other larger species. Water temperature about 24°C.—**Diet:** omnivorous. Some live food is essential if the fish are to breed and come into full coloration.—**Breeding:** relatively difficult, requiring an acid water and a varied diet.

Blood Characin *Hyphessobrycon callistus*

Characidae

Often confused with the very similar serpae tetra (see p. 28), which some authorities classify as a subspecies of *H. callistus*.—**Characteristics:** c. 4cm. The small shoulder spot becomes paler with age. The females are generally somewhat larger than the males. These are active shoaling fishes of the middle and lower water layers which are mostly peaceful, except that they sometimes nibble the fins of other species.—**Distribution:** central South America.—**Maintenance:** very simple in a community tank with clumps of plants. They should be kept as a shoal, old solitary individuals being rather quarrelsome. Water temperature about 24°C.—**Diet:** dried and live food.—**Breeding:** not particularly difficult, but it requires soft, slightly acid water.

The illustration shows the subspecies *H. callistus copelandii*.

Black Neon *Hyphessobrycon herbertaxelrodi*

Characidae

In spite of the popular name this species is not at all that closely related to the neon tetra (p. 34) and it is classified in a different genus.—**Characteristics:** up to 4cm. The males are somewhat smaller and more slender than the females. The characteristic feature of this very peaceful species is the iridescent greenish-black longitudinal band running from the edge of the gill cover to the base of the tail. Black neons like to live as a shoal in the upper and middle water layers.—**Distribution:** South America (Brazil, Rio Taquari).—**Maintenance:** relatively simple, even in a small community tank, with plenty of open water for swimming and some dense vegetation. The iridescent stripe is seen at its best when the lighting is subdued. They must be kept as a shoal. Water temperature about 25°C.—**Diet:** dried food and small live food.—**Breeding:** difficult, and normally only achieved by specialists.

Ornate Tetra *Hyphessobrycon ornatus*

Characidae

This species is frequently confused with the relatively rare *H. rosaceus.*—**Characteristics:** up to 6cm, but usually less. In the males the dorsal fin is much elongated, in the female it is shorter and rounded. These are hardy shoaling fishes which live in the lower water layers.—**Distribution:** South America (Amazon and Guyana).—**Maintenance:** as a shoal in a densely planted community tank. The attractive coloration is enhanced when the substrate is dark. Water temperature about 25°C.—**Diet:** all kinds of dried and live food.—**Breeding:** rather difficult, as some females appear to be infertile.

Neon Tetra *Paracheirodon innesi* (not illustrated)

Characidae

Characteristics: up to 4cm. The iridescent blue band running along each flank from the head to the adipose fin is quite unmistakable. The lower half of the body is an intense red. Both sexes have similar coloration but the males are more slender and have a more concave throat. These are undemanding, shoaling fishes of the middle and lower water layers. They are, however, susceptible to the condition known as neon disease in which the colours become paler. Infected individuals should be caught and killed, in order to prevent the spread of this dangerous disease.—**Distribution:** northern South America (upper Amazon).—**Maintenance:** as a small shoal in any well-planted community tank with fairly subdued light. Young individuals should not be kept with larger species as they may easily be eaten. Water temperature 21-25°C, the optimum being 22°C.—**Diet:** dried food and small live food.—**Breeding:** difficult.

The related cardinal tetra, *Cheirodon axelrodi* (illustrated opposite), has even more brilliant coloration and a completely red underside. It is, however, rather more delicate, even though it is not so susceptible to the dreaded neon disease.

Diamond Tetra *Moenkhausia pittieri*

Characidae

Characteristics: up to 6cm. Sex differences are not very clear. The males are more intensely coloured than the females, the dorsal fin is more elongated and the anal fin rather larger. These are peaceful characins which are mainly active in the upper and middle water layers.—**Distribution:** Venezuela.—**Maintenance:** as a shoal in a large community tank with plenty of space for swimming. Water temperature about 24°C.—**Diet:** plenty of dried and live food. The young require regular amounts of live food for their proper development.—**Breeding:** can be carried out in a small tank, but is not always easy.

Black Tetra *Gymnocorymbus ternetzi*

Characidae

Characteristics: up to 6cm, but usually less. The deep body and strikingly large anal fin are unmistakable. The males are smaller and more slender than the females, and they have white markings on the caudal fin. In both sexes the rear part of the body is deep black, which becomes paler with age. The body also becomes more elongated with the passage of time. This is a peaceful shoaling fish which keeps mainly to the upper and middle water layers.—**Distribution:** South America (Paraguay, Brazil, Bolivia).—**Maintenance:** as a shoal in a well-planted community tank with a large area of open water for swimming. Water temperature about 24°C.—**Diet:** these are greedy omnivores which can be fed almost exclusively on dried food.—**Breeding:** not particularly difficult, and the composition of the water is not very critical.

X-ray Fish *Pristella riddlei*

Characidae

Characteristics: up to 5cm. In spite of its delicate, almost transparent body this is a very robust and hardy aquarium fish. The end of the body cavity, which can be seen externally, tapers posteriorly in the male, but is rounded in the female. These are active, peaceful fishes which form into a tight shoal in the middle water layers.—**Distribution:** South America (Venezuela, Guyana, Amazon).—**Maintenance:** in any community tank with dense vegetation and a large open area for swimming. This species should be kept as a shoal, as solitary individuals or small numbers do not do so well. They thrive best in clear water and in subdued light. Water temperature about 24°C.—**Diet:** dried food and live food.—**Breeding:** relatively simple. It is best to allow them to spawn as a shoal.

Congo Tetra *Micralestes interruptus*

Characidae

Formerly known as *Phenacogrammus interruptus*.—**Characteristics:** up to 12cm, but usually less. A very handsome, rather large African characin with strikingly long fins, particularly the dorsal and anal fins of the males. These are lively, peaceful characins which will form a shoal in the middle water layers.—**Distribution:** central Africa (Zaire).—**Maintenance:** in a large community with scattered clumps of plants and a good area of open water for swimming. In spite of their relatively high price this species should be kept as a shoal which will look at its best if the lighting is not too bright. This is not really a suitable fish for the beginner. Water temperature about 25°C.—**Diet:** plenty of dried and live food. In fact this species must have some live food, preferably insect larvae.—**Breeding:** difficult.

Red-eyed Characin *Arnoldichthys spilopterus*

Characidae

Characteristics: up to 7cm. The strikingly large scales and delicate iridescent coloration are particularly noticeable. The males are more brightly coloured than the females, and they have a red marking on the anal fin, which is lacking in the females. This is a peaceful, hardy but very active shoaling fish which usually swims in the upper and middle water layers.—**Distribution:** West Africa, from Lagos to the Niger delta.—**Maintenance:** as a small shoal in a long, spacious community tank with plenty of swimming space and some dense vegetation. The tank must have a close-fitting lid as the fish tend to jump. Water temperature about 25°C.—**Diet:** dried and live food, but preferably the latter.—**Breeding:** very difficult, and only achieved on a few occasions.

Thayeria boehlkei

Characidae

Characteristics: up to 8cm, but usually somewhat less in the aquarium. A rather inconspicuously coloured fish which typically assumes an oblique position in the water, which is accentuated by the longitudinal black band which extends from the snout to the lower lobe of the caudal fin. The females are generally larger and more rounded than the males. These are peaceful and undemanding shoaling fishes, which live mainly just below the water surface.—**Distribution:** South America (upper Amazon region).—**Maintenance:** as a small shoal in a fairly large community tank with dense vegetation that leaves sufficient open water near the surface. Water temperature about 25°C.—**Diet:** omnivorous. This species can, in fact, be fed almost exclusively on dried food.—**Breeding:** difficult.

Red-spotted Copeina *Copeina guttata*

Characidae

Characteristics: up to 15cm. The colours of this relatively large characin change according to the light. The females are somewhat paler than the males, in which the upper lobe of the caudal fin is elongated.—**Distribution:** central Amazon.— **Maintenance:** as a shoal in a really spacious tank with some large-leaved plants and sufficient open water for swimming. The tank must have a close-fitting lid as these fish tend to jump. Water temperature in the range 15-32°C, but recommended as at least 25°C.—**Diet:** primarily live food, including small insects and insect larvae. Dried food is also taken.— **Breeding:** rather easy, and possible even in hard water.

Blind Cave Characin *Astyanax jordani*

Characidae

Characteristics: up to 7cm. A colourless or pinkish and completely blind fish that lives in caves. Eyes are present in the young but they atrophy as the fish grows older. The adults probably orientate with the help of the lateral line system. The males are somewhat more slender than the females and frequently have black spots on the fins.—**Distribution:** Mexico, in subterranean caves.—**Maintenance:** rather simple. They can be kept in a community tank, but will do better in a small species tank with rocks and a few plants. There is no need for artificial light, although it will not harm the fish, which are interesting to keep. Water temperature about 25°C as in their original habitat, although some authorities recommend lower temperatures.—**Diet:** dried and live food.—**Breeding:** not particularly difficult.

Sumatra Barb, *Barbus tetrazona*

Cyprinidae

Characteristics: up to 7cm, but usually less in the aquarium, where it has long been a popular species, especially suitable for the beginner. There is little difference between the sexes, except that the males usually have more red on the head and ventral fins. These fish tend to nibble the long fins of angelfish and labyrinth fishes, so they are best kept apart.— **Distribution:** Sumatra, Borneo and Thailand.—**Maintenance:** as a large shoal to occupy the middle and lower water layers in a community tank with a large area of open water for swimming. Water temperature about 25°C. If kept too cool the full coloration is not fully developed.—**Diet:** all kinds of dried food, with some greenstuff. Live food will also be appreciated.—**Breeding:** not particularly difficult, providing they are kept as a shoal.

Black Ruby, *Barbus nigrofasciatus*

Cyprinidae

Characteristics: up to 6cm. Superficially rather like the Sumatra barb, but the transverse bands are not so well defined and the head is a characteristic purplish-red. In males at spawning time the front part of the body becomes a particularly handsome crimson colour. The females are always less conspicuously coloured and usually somewhat stouter. These are peaceful, long-lived and active fishes for the middle and lower water layers.—**Distribution:** southern Sri Lanka.—**Maintenance:** in a community tank with plenty of space for swimming. Floating plants are recommended as they slightly reduce the light and this enhances the coloration of the fishes. Water temperature about 25°C.—**Diet:** plenty of dried and live food of all kinds. Live food is not essential but it helps to improve the colours.—**Breeding:** not too difficult.

Schwanenfeld's Barb, *Barbus schwanenfeldi*

Cyprinidae

Characteristics: up to 30cm or more, but usually less in the aquarium. A silvery barb with reddish fins, which in spite of its size is very peaceful; older individuals, however, tend to become aggressive. There are no external differences between the sexes.—**Distribution:** south-east Asia.—**Maintenance:** relatively simple in a large community where the fish have plenty of space for swimming. They should be kept as a shoal but beginners should be warned that young specimens only a few centimetres long grow rapidly and soon become too large for an ordinary community tank. Water temperature about 24°C.—**Diet:** all kinds of dried and live food, with a supplement of greenstuff. In fact, the tank's decorative plants may be attacked.—**Breeding:** not yet achieved in the aquarium, so far as is known.

Green Barb, *Barbus semifasciolatus*

Cyprinidae

Characteristics: c. 7cm. An attractive species with 5-7 dark transverse bands. The females are distinctly stouter, particularly at spawning time, and their colours are somewhat paler than those of the males. These are peaceful, hardy shoaling fishes of the middle and lower water layers.—**Distribution:** south-east China and Hong Kong.—**Maintenance:** in a community tank with scattered plants and, sufficient space for swimming. Water temperature about 22-25°C.—**Diet:** dried and live food.—**Breeding:** not very easy. The fish illustrated opposite and sold as Schubert's barb is probably a selected form of *B. semifasciolatus*.

Zebra Danio, *Brachydanio rerio*

Cyprinidae

Characteristics: up to 6cm. This is an attractive striped fish, particularly suitable for the beginner on account of its hardiness and peaceful nature. The females are somewhat larger and stouter than the males and they have a more convex belly. They should be kept as a shoal which will be active in the upper water layers.—**Distribution:** eastern India.—**Maintenance:** in any community tank, even a small one, provided there is sufficient open water. Over the last decades intensive breeding of this species has tended to reduce the coloration. The tank must be well covered as the fish are likely to jump. Water temperature about 24°C or a little lower.— **Diet:** dried and small live food.—**Breeding:** fairly easy, in almost any type of water. The parents will eat the eggs, so they must be removed immediately after spawning.

Giant Danio, *Danio malabaricus*

Cyprinidae

Characteristics: up to 12cm but usually less. An active shoaling fish which in spite of its fairly large size is peaceful towards the other occupants of the tank. The males are more slender than the females and they have the central blue stripe extending right to the edge of the caudal fin. In the females this stripe turns upwards at the caudal fin. These fish live in the upper water layers.—**Distribution:** Sri Lanka and the west coast of India.—**Maintenance:** always as a small shoal in a large community tank with extensive open water close to the surface. Water temperature about 24°C or slightly lower.—**Diet:** all kinds of dried and live food. If necessary giant danios can be fed exclusively on dried food.—**Breeding:** fairly easy. The parents will attack the eggs so they must be moved to another tank as soon as they have spawned.

Balantiocheilus melanopterus
Cyprinidae

Characteristics: up to 35cm, but usually considerably less in the aquarium. An elegant, slender fish with conspicuously marked fins. The females are usually somewhat stouter than the males. Young individuals are very tolerant, but older specimens tend to be predatory.—**Distribution:** south-east Asia.—**Maintenance:** only in a really spacious tank, but not suitable for a community tank, except when young. The tank must have a good lid as these fish jump. They should be kept as a small shoal. Water temperature about 25°C.—**Diet:** varied, but mainly live food, with a supplement of green-stuff.—**Breeding:** has evidently not yet been achieved.

Elegant Rasbora, *Rasbora elegans*

Cyprinidae

Often recorded in the literature as a subspecies of *R. lateristriata.*—**Characteristics:** up to 13cm. A peaceful fish that can be distinguished by the two dark markings on each flank. The females are paler and somewhat larger and stouter than the males, especially at spawning time. They live in shoals mainly in the upper water layers.—**Distribution:** south-east Asia (Malaya and the Sunda Islands to Borneo).—**Maintenance:** as a shoal in a long, spacious community tank with plenty of space for swimming. Water temperature about 25°C.—**Diet:** dried and live food.—**Breeding:** not particularly difficult.

Harlequin Fish, *Rasbora heteromorpha*

Cyprinidae

Characteristics: up to 4cm. For many years this has been regarded as one of the most suitable fish for the beginner, as it is peaceful and attractively marked. The sexes can be distinguished quite easily. In males the body is more slender and the front edge of the black wedge-shaped marking is rounded, whereas in females it is straight. In addition the males often show territorial behaviour. Harlequins should be kept as a group and they spend most of their time quite close to the surface. They are not aggressive towards other occupants of the tank.—**Distribution:** south-east Asia (Thailand, Malaya and Sumatra).—**Maintenance:** as a large shoal in a spacious community tank. There should be plenty of dense vegetation to provide shelter, but also sufficient open water for swimming. A dark substrate is recommended as this seems to enhance the beautiful golden tints of the scales. They are often kept too cool. The correct temperature for the water is 24-25°C.—**Diet:** a variety of dried and live food, although they will, if necessary, do without the latter.—**Breeding:** not very easy, and there is not much chance of the beginner being successful.

Spotted Rasbora, *Rasbora maculata*

Cyprinidae

Characteristics: up to 2.5cm. One of the smallest aquarium fishes. It is peaceful and very hardy, and lives mainly in small shoals in the upper and middle water layers. The females are somewhat larger and stouter than the males.—**Distribution:** south-east Asia (southern Malaya and Sumatra).—**Maintenance:** as a shoal in a small community tank (c. 40cm long) with dense clumps of fine-leaved plants. They should not be kept with large fishes, otherwise they remain shy and usually hide among the plants. They require a water temperature of at least 24°C.—**Diet:** small dried and live food, the latter, e.g. water-fleas, *Tubifex* and brine shrimp larvae, being essential.—**Breeding:** difficult.

Scissors-tail, *Rasbora trilineata*

Cyprinidae

Characteristics: up to 15cm, but less in the aquarium. Young individuals are almost translucent. The popular name refers to the scissoring movements of the two tail lobes. The females are larger and stouter than the males. These fish live mainly in the upper and middle water layers.—**Distribution:** south-east Asia (Malaya, Sumatra and Borneo).—**Maintenance:** as a shoal, even a small one, in a large community tank with scattered plants and sufficient space for swimming. The tank must have a close-fitting lid as the fish tend to jump. Water temperature about 25°C.—**Diet:** mainly dried food, but live food, especially mosquito larvae, will be appreciated from time to time.—**Breeding:** fairly easy.

White Cloud Mountain Minnow,
Tanichthys albonubes

Cyprinidae

Characteristics: up to 5cm. A decorative little fish that is easy to keep and very suitable for the beginner. They live in small shoals and are very peaceful. The males are smaller and more slender than the females.—**Distribution:** south China, near Canton.—**Maintenance:** in a small, well-planted community tank. The composition of the water is not critical, nor is its temperature which should not, however, exceed 22-23°C. In winter it can even be allowed to fall to 16°C. In fact, this species can be kept in an unheated tank at room temperature.—**Diet:** small dried and live food.—**Breeding:** very easy. Cooler water stimulates spawning. The parents sometimes attack their own eggs.

Red-tailed Labeo, *Labeo bicolor*

Cyprinidae

Characteristics: up to 15cm or more, but usually less in the aquarium. Easily recognized by the velvet-black body and the bright red caudal fin. The sexes are difficult to distinguish, although the females are normally larger than the males. These are hardy fish, but unfortunately intolerant of other members of their own species, sometimes even of other species. They establish territories close to the bottom, which they defend vigorously.—**Distribution:** south-east Asia (Malaya and Thailand).—**Maintenance:** in almost any community tank with hiding-places made from rocks and roots. A single individual will not usually thrive. It is best to keep several young specimens. These will usually form a hierarchy with one fish becoming dominant. Water temperature about 25°C or a little higher.—**Diet:** live and dried food with some greenstuff.—**Breeding:** very difficult, and hitherto not often achieved.

Schultz's Corydoras, *Corydoras schultzei*

Callichthyidae

Characteristics: up to 6.5cm. Distinguished by the iridescent golden band just below the dark brown or blackish back. An uncommonly peaceful, but at the same time active fish which lives near the bottom, using its three pairs of barbels to search for food. The females are somewhat larger and stouter than the males.—**Distribution:** South America (Amazon region). —**Maintenance:** as a small group in a community tank with scattered plants. They like a soft substrate and they get on well with other members of the genus *Corydoras*. Water temperature about 15°C or a little lower.—**Diet:** all kinds of dried and live food. These fishes and their relatives help to keep the tank clean by consuming plant fragments and scraps left by other fishes.—**Breeding:** not particularly difficult.

Corydoras bondi

Callichthyidae

Characteristics: up to 5.5cm. A seldom imported species with two pairs of barbels and a conspicuous dark band extending along each flank to the base of the caudal fin. Here again the females are generally somewhat larger and stouter than the males. These are peaceful and hardy bottom-living fishes.— **Distribution:** northern South America.—**Maintenance:** as a small shoal in a community tank; these are very sociable fishes. Water temperature about 25°C or slightly lower.—**Diet:** dried and live food, with some green-stuff.—**Breeding:** evidently not hitherto achieved.

Blue Corydoras, *Corydoras nattereri*

Callichthyidae

Characteristics: c. 6cm. A pale brown armoured catfish with silvery-blue iridescence and a dark longitudinal band extending from the gill cover to the base of the caudal fin. There are three pairs of barbels. Unfortunately this species is not often available on the market. It is similar to the better known bronze corydoras (*C. aeneus*) which however has only two pairs of barbels. Like other armoured catfishes the blue corydoras is very peaceful and long-lived, spending most of its time on the bottom.—**Distribution:** South America (eastern and central Brazil).—**Maintenance:** as a small group in a community tank. Water temperature about 25°C or a little lower. —**Diet:** all kinds of dried and live food and some greenstuff. —**Breeding:** not very easy, but has been achieved by experienced aquarists.

The enlarged photograph of two armoured catfishes on the opposite page shows the quite characteristic form of the head, and the typical barbels. These are, indeed, among the most popular of all aquarium fishes, not only on account of their rather quaint appearance, but also because they do good work for the aquarist by removing scraps of waste matter from the bottom. They also ensure that small worms which have sunk down from the upper waters do not creep into the substrate where they die and pollute the tank. For this reason it is a good idea to have at least a pair of any of the *Corydoras* species in every community tank. They can also, of course, be kept by themselves in a species tank with a large area of soft sandy substrate and fairly shallow water.

Corydoras grafi

Callichthyidae

Characteristics: up to 6.5cm. This is another rather uncommon armoured catfish, with finely dotted scales and two pairs of barbels. There are no external differences between the sexes, although the females are normally larger and somewhat stouter. Like its relatives this is a peaceful, but sometimes very active bottom-living fish.—**Distribution:** South America (Brazil).—**Maintenance:** as a small group in any community tank. Water temperature about 25°C or a little lower.—**Diet:** all kinds of dried and live food, with some plant matter.—**Breeding:** possible, but not very easy.

Myers's Corydoras, *Corydoras myersi*

Callichthyidae

This species is sometimes described under the name *C. rabauti.*—**Characteristics:** up to 6cm. A handsome armoured catfish with an iridescent reddish-golden body and three pairs of barbels. The females are again larger and usually somewhat stouter than the males. These are peaceful, sociable and active bottom-living fishes.—**Distribution:** South America (Amazon region).—**Maintenance:** as a small group in any community tank. Water temperature about 25°C or a little lower.—**Diet:** dried and live food with some plant fragments.—**Breeding:** not very easy, but has been successful in the hands of experienced aquarists.

Dwarf Corydoras, *Corydoras hastatus*

Callichthyidae

Characteristics: up to 3.5cm. Very small and comparatively slender armoured catfishes. Their habits differ somewhat from those of related species, because they spend more time swimming and are often seen resting on a leaf in the upper water layers. They are just as peaceful and sociable as their relatives.—**Distribution:** South America (Amazon basin).— **Maintenance:** in any kind of community tank, even the smallest. They should certainly be kept as a small group. Water temperature about 25°C or somewhat lower.—**Diet:** dried and live food and some greenstuff.—**Breeding:** not particularly difficult.

Peppered Corydoras, *Corydoras paleatus*

Callichthyidae

Characteristics: up to 7cm. This very undemanding, long-lived species is one of the best known of the armoured catfishes. The iridescent body is marked with irregular dark spots and fine dots. The males are more slender than the females and have a more pointed dorsal fin. There is also a selected form with a flesh-coloured body and red eyes which is just as easy to keep.—**Distribution:** south-east Brazil and La Plata basin.—**Maintenance:** as a small group in a community tank which can be quite small. The composition of the water is not critical. Water temperature about 25°C, or it can be allowed to fall to 18°C.—**Diet:** dried and live food, in addition to scraps picked up from the bottom.—**Breeding:** very easy, and even possible in a community tank. Lower temperatures stimulate spawning.

Reticulated Corydoras, *Corydoras reticulatus*

Callichthyidae

Characteristics: up to 7cm. One of the most beautiful armoured catfishes with dark reticulation on an iridescent greenish-red background; there are two pairs of barbels. The females are mostly larger and somewhat stouter than the males, and their colours are not so contrasting. These are very active, bottom-living fishes which are no trouble to keep.—**Distribution:** South America (Amazon basin).—**Maintenance:** as a small group in a community tank furnished with a couple of smooth flat rocks on which the fish like to rest. Water temperature about 25°C or less.—**Diet:** dried and live food, with a supplement of greenstuff.—**Breeding:** not easy, but it has been achieved.

Short-bodied Catfish, *Brochis coeruleus*

Callichthyidae

Characteristics: up to 8cm. Although classified in a separate genus this peaceful and long-lived species is very similar in appearance, behaviour and requirements to the armoured catfishes already described. The iridescent, greenish body is relatively tall, and there are three pairs of barbels. The sexes are very difficult to distinguish. These are sociable, bottom-living fishes which also swim about in the upper waters.—**Distribution:** South America (upper Amazon).—**Maintenance:** as a group, possibly with other armoured catfishes, in a well-planted community tank with hiding-places and sufficient space for swimming.—**Diet:** all kinds of dried and live food, with supplementary boiled lettuce as well as waste scraps from the bottom.—**Breeding:** not very easy, but it has been achieved.

Lyretail, *Aphyosemion australe*

Cyprinodontidae

Characteristics: up to 6cm. A very colourful, but unfortunately short-lived fish; this also applies to its relatives. The females are rather less striking and lack the prolongations of the dorsal and caudal fins which are so characteristic of the males. Lyretails swim mainly in the middle and lower water layers.—**Distribution:** West Africa, in coastal swamps of Gabon and Cameroun.—**Maintenance:** like most of the egg-laying toothcarps this is not a suitable fish for a community tank, although it can sometimes be kept with other species of *Aphyosemion*. It is best to keep one male and one or more females in a small, densely planted species tank with a dark substrate. Water temperature 22-24°C.—**Diet:** mainly live food with a supplement of dried food.—**Breeding:** not too difficult for the advanced aquarist.

Red Lyretail, *Aphyosemion bivittatum*

Cyprinodontidae

Characteristics: up to 6cm. A variably coloured species, that evidently comprises several, slightly differing local populations. The females have dull colours and rounded fins. These are attractive and very peaceful fish that swim around in the upper and middle water layers.—**Distribution:** West Africa, from Togo to Cameroun.—**Maintenance:** is possible in a densely planted community tank, but it is much better to keep them in a small species tank with dense clumps of plants and a dark substrate. Water temperature: 22-24°C.—**Diet:** mainly live food in variety, but dried food will also be taken. —**Breeding:** not too difficult for the experienced aquarist.

Christy's Lyretail, *Aphyosemion christyi*

Cyprinodontidae

This species is sometimes described under the name *A. schoutedeni.*—**Characteristics:** c. 5cm. A very variable, but always brightly coloured species that was first introduced into Europe at the end of the 1940s. Its systematic position is still not quite clear, and there may be confusion between it and closely related and very similar species. The females are duller than the males.—**Distribution:** central Africa (Congo region).—**Maintenance:** as a pair in a densely planted community tank, possibly together with other *Aphyosemion* species, or even better in a small species tank.—**Diet:** live food, with a supplement of dried food.—**Breeding:** not too difficult. The different colour varieties and geographical races can be crossed with one another, but the offspring are often infertile.

Aphyosemion bualanum

Cyprinodontidae

Characteristics: up to 5cm. Not so frequently seen in the home aquarium, but a very desirable species. It has no special requirements as regards the composition of the water. The females are duller than the males and their fins are not pointed.—**Distribution:** central Africa, east Cameroun and the Central African Republic.—**Maintenance:** is possible in a community tank, but better as a pair in a small species tank with dense vegetation and a dark substrate. Water temperature about 24°C.—**Diet:** mainly live food, with some dried food.—**Breeding:** not particularly difficult.

Plumed Lyretail, *Aphyosemion filamentosum*

Cyprinodontidae

Characteristics: up to 5.5cm. A rather variable species with a relatively short body, and an elaborate caudal fin in the male. The female is much shorter (c. 3cm) and duller, with rounded fins. These fish which live mostly in the middle water layers are more active than most other species of *Aphyosemion.*— **Distribution:** West Africa (swamps from south-western Nigeria to Cameroun).—**Maintenance:** in a small species tank with some dense vegetation, a dark substrate and sufficient open water for swimming. The species is scarcely suitable for a community tank. Water temperature 20-24°C.—**Diet:** mainly live food, with some dried food.—**Breeding:** relatively difficult.

Aphyosemion cognatum

Cyprinodontidae

Characteristics: up to 6cm. The body has numerous red markings, and the fins of the male are not so elongated as in most other members of the genus. The female is smaller and duller with rounded fins. This is a somewhat difficult fish for the aquarium and is not really suitable for the beginner. It lives mostly in the upper water layers.—**Distribution:** Africa (lower Congo).—**Maintenance:** as a pair, possibly with other species of *Aphyosemion,* in a small tank with a dark substrate and dense vegetation. Water temperature 20-24°C.—**Diet:** mainly live food in variety, with some dried food.—**Breeding:** not very easy.

Blue Gularis, *Aphyosemion sjoestedti*

Cyprinodontidae

The name *A. sjoestedti* has also been used for the fish that is nowadays known as *Roloffia occidentalis* (see p. 84).— **Characteristics:** up to 12cm, but usually smaller in the aquarium. One of the larger members of the genus, distinguished by the three points to the caudal fin of the male. The females are smaller and less conspicuously marked, with more rounded fins. In general a very quarrelsome species.— **Distribution:** West Africa (southern Nigeria to Cameroun). —**Maintenance:** in a well-planted but not too small species tank. Not suitable for keeping in a community tank with smaller species, owing to its predatory propensities. Water temperature 20-24°C.—**Diet:** live food with a supplement of dried food.—**Breeding:** not very easy.

Steel-blue Aphyosemion,
Aphyosemion gardneri

Cyprinodontidae

Formerly known as *A. nigerianum.*—**Characteristics:** up to 6cm. The coloration is very beautiful but variable. In some males the fins, but not the pectoral fins, have bright golden-yellow edges. The females are somewhat smaller with duller coloration. An aggressive fish.—**Distribution:** West Africa (Nigeria and western Cameroun).—**Maintenance:** not suitable for a community tank on account of their aggressive habits. Best kept in a small species tank with suitable hiding-places in which the females can seek shelter from the male. It is, in fact, recommended that one male should be kept with several females, as a single female would soon become exhausted by the aggressive behaviour of the male.—**Diet:** mainly live food, with some dried food.—**Breeding:** relatively easy.

Blue Panchax, *Aplocheilus panchax*

Cyprinodontidae

Characteristics: up to 8cm. The body is elongated and pike-like with the dorsal fin situated far to the rear. Externally the sexes are not very distinct, but the female is normally somewhat paler. This is a quiet, rather shy fish which can at times become aggressive and is then a danger to other fishes. It lives mainly in the upper water layers.—**Distribution:** south-east Asia (India to Indonesia).—**Maintenance:** either in a community tank with larger fishes or in a small, not too tall species tank, which must be densely planted and also well covered to prevent the fish from jumping out. Otherwise these are hardy fish and quite easy to keep. Water temperature about 24°C or a little lower.—**Diet:** live and dried food.— **Breeding:** fairly easy, and even possible in a community tank.

Ceylon Killifish, *Aplocheilus dayi*

Cyprinodontidae

Characteristics: up to 9cm. A brightly coloured and robust species which can be recommended for beginners. The sexes are often difficult to distinguish, although the females are usually somewhat paler and have a more rounded caudal fin. In spite of their apparent shyness these are aggressive predators.—**Distribution:** south-east Asia (Sri Lanka).— **Maintenance:** in a community tank with dense vegetation, some hiding-places and a close-fitting lid. On no account should this species be kept with small, delicate fishes. Water temperature about 25°C.—**Diet:** live food is preferred, but they will take dead food.—**Breeding:** not particularly difficult.

Chaper's Epiplatys, *Epiplatys chaperi*

Cyprinodontidae

Characteristics: up to 6cm. This is the true *E. chaperi* (see next species). The females have duller colours and more rounded fins than the males. This is a typical surface-living fish which is very peaceful.—**Distribution:** West Africa (Ghana and Ivory Coast).—**Maintenance:** is possible in a community tank, but it is preferable to keep these fish in a well-planted species tank with sufficient hiding-places. It is advisable to change a proportion of the water at frequent intervals. Water temperature 20-24°C.—**Diet:** live and dried food in variety, and particularly insect larvae.—**Breeding:** not too difficult.

Daget's Epiplatys, *Epiplatys dageti*

Cyprinodontidae

For many years this fish was erroneously known as *E. chaperi*.—**Characteristics:** up to 6cm. A slender, pike-like fish with 5-6 dark transverse bands. The females are somewhat smaller than the males and they have rounded fins. In the subspecies *E. dageti monroviae*, which is the form usually kept in the aquarium, the male is also distinguished by having a red throat. These are robust but short-lived fishes which spend most of their time close to the surface.—**Distribution:** West Africa (Liberia to Ghana).—**Maintenance:** in a densely planted community tank with open water for swimming, but not with smaller fishes. Water temperature about 24°C.—**Diet:** plenty of live and dried food.—**Breeding:** not too difficult, but rearing the young requires great care.

Six-barred Epiplatys, *Epiplatys sexfasciatus*

Cyprinodontidae

Characteristics: up to 10cm. A rather large species with iridescent greenish-yellow scales and usually 6 dark bars below the lateral line. The females are smaller and duller than the males and their fins are more rounded. This is a predatory, surface-living species which will attack smaller fishes.—**Distribution:** West Africa (Liberia to Congo). —**Maintenance:** in a well-covered species tank, that is not too small, with some dense vegetation and space for swimming. Part of the surface can be covered with floating plants. This species can be kept in a community tank provided the other occupants are the same size or larger. Water temperature about 25°C.—**Diet:** plenty of food, including small fish. Dried food is also taken.—**Breeding:** rather difficult.

Lamotte's Epiplatys, *Epiplatys lamottei*

Cyprinodontidae

Characteristics: up to 7cm. A recently described species which has not been known for long in the aquarium world. The females are brownish with a slight violet tinge, the males have a pattern of intensive bluish-violet markings. In contrast to most of its relatives this is not a typical surface-living fish.—**Distribution:** West Africa (Guinea to Liberia).—**Maintenance:** in a well-planted species tank with subdued lighting which enhances the coloration. Water temperature 21-23°C.—**Diet:** live food in variety. Dried food can also be offered and they may become accustomed to it. Evidently in the wild they feed primarily on ants.—**Breeding:** rather difficult.

Guenther's Nothobranch,

Nothobranchius guentheri

Cyprinodontidae

Characteristics: c. 6cm. A very handsome but rather delicate fish, which does not live long, even in the wild. The females are smaller with paler coloration.—**Distribution:** East Africa, coastal areas of Kenya and Tanzania.—**Maintenance:** as a pair in a medium-sized but not too tall species tank with scattered clumps of plants, including some that float. The males do not tolerate one another, and the species as a whole is too aggressive for a community tank. The substrate should preferably be of soft sand. Water temperature 20-25°C.—**Diet:** mainly live food, but dried food will also be eaten.—**Breeding:** not too difficult.

Rachow's Nothobranch,

Nothobranchius rachovii

Cyprinodontidae

Characteristics: up to 6cm. A rather shy species which can become very intolerant at times. The males in particular quarrel among themselves. Another disadvantage is the short life span, which is usually less than a year. Only the males have the brilliant coloration shown here, the females are an inconspicuous pale brown.—**Distribution:** south-eastern Africa, coastal areas of Mozambique.—**Maintenance:** as a pair in a shallow species tank with scattered plants and a soft substrate. This species can only be recommended for experienced aquarists. Water temperature 20-25°C.—**Diet:** abundant live food of all types, with some dried food as a supplement.—**Breeding:** not too difficult.

Playfair's Panchax, *Pachypanchax playfairi*

Cyprinodontidae

Characteristics: up to 10cm. A handsome fish with the typical pike-like shape, which is very robust and lively but unfortunately, like almost all its relatives, also rather aggressive. It can, however, be recommended for the beginner. The somewhat smaller females can be recognized by their less conspicuous colours and fins.—**Distribution:** East Africa (coastal areas of Seychelles, Madagascar and Zanzibar).—**Maintenance:** is possible in a spacious, well-planted community tank with other fishes of the same size, but is better in a species tank with clumps of plants and some floating plants. Water temperature about 24°C, water composition not critical.—**Diet:** plenty of live food, including small fish, with some dried food.—**Breeding:** not particularly difficult.

Red Aphyosemion, *Roloffia occidentalis*

Cyprinodontidae

This species was formerly included under *Aphyosemion sjoestedti* (see p.74).—**Characteristics:** up to 9cm. The males are rather variable but always colourful. The females show much duller coloration and have a rounded caudal fin, whereas that of the males is fan-shaped. A suitable fish for the beginner, it is hardy and normally fairly peaceful, but it can become aggressive and predatory.—**Distribution:** West Africa (Sierra Leone).—**Maintenance:** is possible in a community tank with other species of the same size, but it is preferable to keep a pair in a species tank with scattered plants and a dark substrate. Water temperature 22-24°C.—**Diet:** plenty of live food of all kinds. Dried food is usually refused.—**Breeding:** not too difficult.

Black Molly, *Poecilia sphenops* (domesticated form)

Poeciliidae

Formerly known as *Mollienesia sphenops*. **Characteristics:** c. 6cm. The black molly is the best-known domesticated form of the pointed-mouth molly *P. sphenops*. The body is rather squat and uniformly velvety-black. The females are larger and stouter than the males. These are peaceful fish which swim about in the upper water layers. Other domesticated forms include the liberty black molly, the Crescenty black molly and Sternke's black Yucatan.—**Distribution:** originally central America, from Venezuela through Colombia and Mexico to Texas; also in the Leeward Islands.—**Maintenance:** in a community tank, with at least a pair. Only the standard black molly should be kept by a beginner as some of the other forms are rather delicate. Water temperature about 25°C or a little higher.—**Diet:** dried and live food, but a supplement of greenstuff is even more important than live food.—**Breeding:** not difficult.

Sailfin Molly, *Poecilia velifera*

Poeciliidae

Characteristics: up to 15cm, but usually less in the aquarium. A species that is easily distinguished by the very tall sail-like dorsal fin. The females are somewhat larger and stouter than the males and generally not so intensely coloured; also they usually have a smaller dorsal fin. These are very peaceful, hardy fish which live mainly in the upper water layers.—**Distribution:** southern Mexico (coastal zone of Yucatan).—**Maintenance:** as a pair in a spacious, well-planted community tank with plenty of open water for swimming. The tank must be well covered as these fish jump. It is best to keep only a pair together, as the males tend to be intolerant of one another. Water temperature about 25°C or a little higher.—**Diet:** dried and live food, and greenstuff is essential. The fish browse on algae.—**Breeding:** not very difficult.

Guppy, *Poecilia reticulata*

Poeciliidae

Formerly known as *Lebistes reticulatus.*—**Characteristics:** 3-4cm for males; the inconspicuous females are almost twice this length. In their natural habitat guppies are widely distributed and very variable in form and coloration. They are also very prolific. It is not surprising, therefore, that since the species was first described in 1859 aquarists have bred and selected a large number of varieties. These vary principally in coloration and fin shape. The photographs on the opposite page show two of these varieties. For decades the guppy has been regarded as an ideal fish for the beginner, although some of the highly selected forms are rather delicate. In general, guppies are active, undemanding, peaceful and hardy. They live mainly in the upper water layers.—**Distribution:** South America (north of the Amazon, Guyana, Venezuela, Trinidad and Barbados).—**Maintenance:** in almost any community tank with a length of 40cm or more. There should be some dense vegetation and sufficient space so that the large fins can be allowed to develop properly. The composition of the water is not critical, and its temperature can be in the range 20-28°C, with an optimum at 22-24°C.—**Diet:** all kinds of dried and live food with some greenstuff.—**Breeding:** very easy, and in fact it usually occurs without help from the aquarist. A female can give birth to young every four weeks.

Swordtail, *Xiphophorus helleri*

Poeciliidae

Characteristics: males are up to 15cm long, including the elongated lower caudal fin lobe, while the much stouter females which lack this feature reach a length of c. 12cm. The green swordtail, which occurs in the wild in a number of slightly different colour forms is a basic type from which aquarists have over a period of years produced the various domesticated forms now available. Hybridization with related members of the genus *Xiphophorus* has also produced numerous varieties, including the red swordtail (uniformly red), the yellow swordtail, the red wagtail swordtail (red with black fins), the Berlin hybrid swordtail (yellowish-red marked with black spots), the Wiesbaden cross (double sword) and also albino swordtails. Most of these forms are peaceful and undemanding and very suitable for the beginner. The males are sometimes quarrelsome and a hierarchy or pecking order may develop in the tank. This species also shows sex change: an old female may develop into a male, complete with sword.—**Distribution:** central America (Mexico, Honduras and Guatemala).—**Maintenance:** is possible in any community tank, with dense vegetation and sufficient space for swimming, particularly in the upper part of the tank. Swordtails can thrive at any temperature between 17°C and 26°C, but the optimum is 24°C. The composition of the water is not critical.—**Diet:** dried and live food with some greenstuff and algae.—**Breeding:** quite easy, even for the beginner.

The upper photograph shows a form with typical dark markings, the lower photograph shows the popular red domesticated swordtail.

Platy, *Xiphophorus maculatus*

Poeciliidae

Formerly known as *Platypoecilus maculatus.*—**Characteristics:** males up to 4cm, females up to 6cm long. This species is also variable in form and coloration in the wild, and so it is not surprising that over the years aquarists have developed a large number of domesticated forms of the platy. One of the most popular is the red platy which is brilliant red from snout to tail. Then there are yellow platies, golden platies, wagtail platies (red with black fins), tuxedo platies (red or green with black flanks) and several chequered or spotted varieties. In particular, there are numerous forms resulting from the hybridization of platies and swordtails, although some of these are sterile. With the exception of the highly domesticated forms all platies are peaceful, hardy and undemanding fishes which are ideal for the beginner. They spend most of their time in the upper water layers.—**Distribution:** Central and South America (Mexico, Guatemala).—**Maintenance:** in any community tank, even a small one, with dense clumps of plants and open water for swimming. These are sociable fishes which should be kept as a shoal. Water temperature 20-25°C.—**Diet:** all kinds of dried and live food, and some greenstuff is essential. Algae are also eaten.—**Breeding:** fairly easy, even for the beginner.

Ramirez's Dwarf Cichlid,

Apistogramma ramirezi

Cichlidae

The systematic position of this cichlid is somewhat uncertain. It is sometimes placed in the genus *Microgeophagus*. —**Characteristics:** up to 6cm. The females are generally smaller than the males, otherwise there are no external sex differences. This is a very peaceful and sociable cichlid, but unfortunately it only lives for about two years, spending most of its time close to the bottom.—**Distribution:** South America (Venezuela).—**Maintenance:** as a pair in a community or species tank which should be well planted and have a number of cave-like hiding-places made from rocks and roots or old coconut shells. The water should not be too hard, and the optimum temperature is about 24°C.—**Diet:** mainly live food, but dried food, including freeze-dried, will be taken. —**Breeding:** not too difficult.

Borelli's Dwarf Cichlid, *Apistogramma borellii*

Cichlidae

Characteristics: up to 8cm. The inconspicuously coloured females are about 3cm shorter. The coloration of the males varies from yellow to grey. The broad, dark longitudinal stripe varies in prominence. These are hardy cichlids which are particularly handsome at spawning time, but unfortunately they are very intolerant of other species. Like their relatives they live mostly in the lower water layers.— **Distribution:** South America (southern Brazil to the Rio Paraguay).—**Maintenance:** not really suitable for a community tank on account of their aggressive behaviour, so they should be kept in a species tank that is not too small, with plants and plenty of hiding-places. Water temperature about 24°C.—**Diet:** live food with some dried food, including freeze-dried water-fleas.—**Breeding:** rather difficult.

Yellow Dwarf Cichlid, *Apistogramma reitzigi*

Cichlidae

Characteristics: up to 8cm. The females are smaller and less brightly coloured than the males. These are peaceful, undemanding little cichlids which live near the bottom, where they establish small territories.—**Distribution:** South America (basin of the Rio Paraguay).—**Maintenance:** in almost any community tank in which they alone occupy the lower water lazers. There should be several dense clumps of water plants, as well as hiding-places, a dark substrate and a reasonable amount of open water for swimming. There should be only a single pair in a normal tank, as the males tend to quarrel with one another. Water temperature about 24°C.—**Diet:** live food with some dried food.—**Breeding:** not very difficult.

Dimidiatus, *Nannochromis dimidiatus*

Cichlidae

Characteristics: up to 8cm. The females are smaller, less brightly coloured and somewhat more rounded than the males which are mainly reddish-brown. This species is not often available on the market, which is a pity because it is peaceful and shows interesting behaviour patterns. They live almost entirely on or near the bottom and like to hide in cavities which they defend as territories against all intruders.—**Distribution:** Africa (Zaire and the Central African Republic).—**Maintenance:** as a pair in a well-planted community tank, provided there are sufficient hiding-places and there is no disturbance by other bottom-living fishes. Alternatively, the pair can be kept in a species tank which need not be very large. It is important to renew a proportion of the water at regular intervals. Water temperature about 25°C.—**Diet:** live and dried food.—**Breeding:** not very easy.

Keyhole Cichlid, *Aequidens maronii*

Cichlidae

Characteristics: males up to 15cm, but usually much less in the aquarium, females smaller with the fins less elongated. A variable and not particularly beautiful species, but one of the most popular cichlids because it is peaceful and easy to keep. It lives near the bottom where it establishes a territory that must not be disturbed by other fishes.—**Distribution:** north-eastern South America (Guyana).—**Maintenance:** in a community tank with plants and hiding-places. Water temperature about 25°C.—**Diet:** dried and live food. In fact these easily satisfied fish can be fed mainly on dried and freeze-dried food.—**Breeding:** relatively easy, and possible even for the beginner.

Flag Cichlid, *Aequidens curviceps*

Cichlidae

Characteristics: up to 8cm. The females are somewhat smaller and paler than the males and their fins are not so well developed. These are peaceful, hardy fishes that are very suitable for the beginner. Like their relatives they establish territories on the bottom, but unlike most cichlids they do not attack the plants, even those with delicate leaves.—**Distribution:** South America (Amazon basin).—**Maintenance:** in a well-planted community tank with hiding-places and open water for swimming. The composition of the water is not critical, but it is advisable to change a proportion of it at frequent intervals. Water temperature about 24°C.—**Diet:** omnivorous, and in fact these fish can be fed largely on freeze-dried and ordinary dried food, but they require some live food.—**Breeding:** rather easy, the only difficulty being to find a compatible pair.

Firemouth Cichlid, *Cichlasoma meeki*

Cichlidae

Characteristics: up to 15cm, but usually less. the females are somewhat smaller and paler. At spawning time the males are particularly colourful. These are active territorial fishes which, in spite of their size, are quite peaceful. Unfortunately they dig and disarrange the substrate and plants.— **Distribution:** Central America (Mexico and Guatemala). —**Maintenance:** is very easy, in a spacious community tank, but only with other fishes of a similar size. Adults are best kept as a pair in a fairly large species tank with a coarse gravel substrate and tough plants. Water temperature about 24°C.—**Diet:** mainly dried food with some live food. —**Breeding:** fairly easy.

Banded Cichlid, *Cichlasoma severum*

Cichlidae

Characteristics: up to 20cm, but usually much less in the aquarium. The coloration is variable and the sexes are difficult to distinguish. The males generally have more elongated fins, but this is not an entirely reliable character. These are hardy, peaceful fish which may become rather aggressive at spawning time. They live near the bottom and establish territories.—**Distribution:** South America (Guyana and the northern Amazon basin).—**Maintenance:** is possible in a community tank, provided the sexes are kept separate and only put together in a breeding tank when ready to spawn. They must be supplied with hiding-places.—**Diet:** live and dried food with some greenstuff. They will eat earthworms, mealworms and finely chopped meat (heart).—**Breeding:** very difficult.

Lamprologus brichardi

Cichlidae

This is a rarely imported cichlid species formerly known as *L. savoryi elongatus*.—**Characteristics:** c. 7cm. In this attractive species it is not possible to distinguish the sexes externally. To obtain a pair for breeding it would be necessary to put several individuals in a tank and allow them to sort themselves out. They establish territories on the bottom, which they defend vigorously.—**Distribution:** only in Lake Tanganyika.—**Maintenance:** in a community tank but only with large species which live in the upper water layers. Otherwise they should be kept on their own. They require plenty of places in which to hide. Water temperature about 25°C.—**Diet:** mainly live food in variety.—**Breeding:** relatively difficult.

Pelvicachromis pulcher

Cichlidae

Formerly known as *Pelmatochromis kribensis*.—**Characteristics:** males up to 9cm, but the females are somewhat smaller with rounded dorsal and anal fins. Both sexes are brightly coloured, but rather variable. This is a very peaceful, territorial, bottom-living fish which is suitable for the beginner.—**Distribution:** West Africa (Niger delta).—**Maintenance:** in a community tank with clumps of plants and hiding-places, but only with robust fishes that keep to the upper waters. It is best to keep only a single pair in the tank. Water temperature about 24°C.—**Diet:** live and dried food.—**Breeding:** fairly easy, and sometimes possible even in a community tank.

Deep Angelfish, *Pterophyllum altum*

Cichlidae

Characteristics: the angelfishes of the genus *Pterophyllum* are all rather similar in appearance, and can only be distinguished after careful examination. They all have a tall, disc-shaped body and very long dorsal and anal fins. The species shown here, which is unfortunately not often available on the market, has extremely long fins. It is difficult to give an exact idea of the height as this varies according to the conditions. Some authors speak of 40-50cm, but it is always less than this in the aquarium. The sexes cannot be distinguished externally, except at spawning time. This is an expensive species, and so only suitable for the experienced aquarist. Deep angelfishes are quiet and live mainly in the middle water layers, but they react rapidly to sudden disturbances.—**Distribution:** South America (Orinoco).—**Maintenance:** in a spacious community tank but only with larger, peaceful species. Small fishes, such as neon tetras, will quickly be eaten. The tank should be planted with dense clumps of *Vallisneria,* leaving sufficient space in the centre for swimming. Several individuals should be kept together. Water temperature about 24°C.—**Diet:** various live foods, with some dried food and greenstuff.—**Breeding:** very difficult and not often achieved.

Angelfish, *Pterophyllum scalare*

Cichlidae

Characteristics: up to 15cm long and 25cm tall, including the fins, but less in the home aquarium. The sexes cannot be distinguished externally, except at spawning time. These are extremely popular, long-lived fishes which swim in the middle water layers, moving around quite slowly in shoals. There are now several domesticated forms, such as the black-lace angelfish, the marbled angelfish and the recently developed golden angelfish shown opposite.—**Distribution:** South America (Amazon).—**Maintenance:** in a large, talw community tank with clumps of plants, such as *Vallisneria*, to provide shelter. Angelfishes should only be kept with larger, peaceful species. Water temperature about 25°C, but slightly higher for the more delicate domesticated forms.—**Diet:** live and dried food with some greenstuff. Do not overfeed. —**Breeding:** not very easy.

Discus, *Symphysodon discus*

Cichlidae

Characteristics: up to 20cm, but usually less. The sexes are difficult to distinguish externally. This species is characterized by the conspicuous dark vertical bands, which are lacking in the related *Symphysodon aequifasciata* which has three subspecies. Discus fishes are peaceful and long-lived, but are only suitable for the experienced aquarist as they have special requirements as regards water composition and general care. They also require more warmth than most aquarium fishes. —**Distribution:** South America (middle Amazon).—**Maintenance:** in a large species tank, approximately 100cm long, with tall, tough plants. The substrate should be dark and there should be rocks and roots arranged to provide hiding-places. Water temperature 25-30°C.—**Diet:** mainly live food in variety, with some dried food and greenstuff. Some aquarists believe that live food should not be fed.—**Breeding:** very difficult, and only for the specialist.

Pearl Gourami, *Trichogaster leeri*

Anabantidae

Characteristics: up to 15cm, but usually less in the aquarium. Perhaps the most attractive of the Anabantidae. During the spawning periods the flanks of the males are covered with a mosaic of iridescent spots. The dorsal and anal fins are much elongated and the ventral fins, which are situated far forwards, are modified to form long filaments. In the females the colours are less brilliant and the dorsal fin does not end in a point. These are very peaceful but rather shy fishes which may be somewhat delicate. They are, however, quite suitable for the beginner.—**Distribution:** south-east Asia (Thailand, Malaya, Sumatra, Borneo).—**Maintenance:** as a pair or a shoal in a spacious, well-planted community tank with fairly shallow water. They should not be put in with very temperamental fishes, nor with those such as Sumatra barbs which will nibble their long fins. Water temperature about 25°C or more.—**Diet:** dried and live food with some greenstuff.—**Breeding:** very difficult.

Three-spot Gourami, *Trichogaster trichopterus*

Anabantidae

Characteristics: up to 15cm. The males have a taller, more pointed and elongated dorsal fin than the females. From this relatively inconspicuous species breeders have developed the marbled Cosby form *T. trichopterus* 'Cosby' and the blue gourami *T. trichopterus 'sumatranus'* shown here. These are not true subspecies but domesticated forms. They are all peaceful, active and easy to keep.—**Distribution:** south-east Asia (Thailand, Malaya, Sumatra, Java, Borneo).— **Maintenance:** in any well-planted community tank with open water for swimming. Water temperature about 24°C. —**Diet:** all kinds of dried and live food.—**Breeding:** relatively easy.

Moonlight Gourami, *Trichogaster microlepis*

Anabantidae

Characteristics: up to 15cm, but usually less. On account of the fine scales the flanks show a delicate silky fluorescence. The sexes are difficult to distinguish. Males have a somewhat larger and more pointed dorsal fin and their filamentous ventral fins are orange, whereas in the females they are more yellowish. These are peaceful, often rather shy fishes which live mostly in the middle water layers.—**Distribution:** south-east Asia (Cambodia and Thailand).—**Maintenance:** in a spacious community tank with clumps of plants to provide shelter and open water for swimming, but not with restless fishes or those that are too small. Water temperature about 25°C or slightly higher.—**Diet:** all kinds of dried and live food.—**Breeding:** not very easy.

Chocolate Gourami,
Sphaerichthys osphromenoides

Anabantidae

Characteristics: up to 7cm. This is an attractive chocolate-brown species that is certainly peaceful, but it is very delicate and susceptible to disease. It is therefore only suitable for the experienced aquarist. The sexes are difficult to distinguish. The females usually have narrow dorsal and anal fins, the dorsal being less pointed. These are delicate fishes which always remain shy. When alarmed or in poor condition the coloration becomes paler.—**Distribution:** south-east Asia (Malaya and Sumatra).—**Maintenance:** as a pair or a small shoal in a large species tank with fairly shallow water and dense vegetation. The water must be soft, slightly acid and preferably filtered through peat. Water temperature 26-30°C.—**Diet:** mainly live food in variety. Only exceptionally will they take dried food.—**Breeding:** very difficult and only achieved on a few occasions. The eggs are brooded in the mouth of one of the parents

Siamese Fighting Fish *Betta splendens*

Anabantidae

Characteristics: up to 6cm. Only the males have the much enlarged fins. The females are smaller with duller coloration. The original wild form of this species is very variable, and over a long period of years breeders have succeeded in producing domesticated races with large fins and a variety of colours. The plate shows just two of these varieties. The popular name refers to the extremely aggressive behaviour of the males. When confined in a small space two males will fight until their fins are torn to shreds, and such fights are often fatal. On the other hand, they are very tolerant of other species. In Thailand special fights between rival males have been staged. Fighting fishes have a fairly short life, on average about two years.—**Distribution:** south-east Asia (Thailand, Malaya, Singapore).—**Maintenance:** is possible in a community tank but preferable in a small species tank with either a pair or one male and several females. The tank should have some dense vegetation and it should not be too tall. The aggressive behaviour of the males can be observed by putting two in a tank and separating them by a glass pane. The mere sight of a rival is sufficient to release the full aggressive behaviour pattern. Water temperature at least 25°C.—**Diet:** all kinds of live and dried food.—**Breeding:** quite easy. Fighting fishes show very interesting courtship and parental care. Like most of the other labyrinth fishes the male builds a nest of bubbles at the surface which holds the eggs until they hatch.

Thick-lipped Gourami, *Colisa labiosa*

Anabantidae

Characteristics: up to 10cm. This species can be recognized by the strikingly thick lips. It is, however, very similar to the banded gourami, *C. fasciata*. The males can be distinguished by the elongated and pointed dorsal and anal fins and by their more intense coloration. These are peaceful fishes which live mainly in the middle water layers.—**Distribution:** south-east Asia (southern Burma).—**Maintenance:** as a pair or small shoal in a large community tank with dense clumps of plants and open water in the middle for swimming. They should not be kept with restless or quarrelsome species. Water temperature about 25°C; at lower temperatures the full coloration is not developed.—**Diet:** all kinds of dried and live food, with some greenstuff.—**Breeding:** not particularly difficult.

Dwarf Gourami, *Colisa lalia*

Anabantidae

Characteristics: up to 6cm. A smaller and more squat fish than those described above. The males have brighter colours and a more marked pattern than the females. These are peaceful and undemanding fishes which swim about in the middle water layers.—**Distribution:** south-east Asia (Bengal and Assam).—**Maintenance:** in quite a small community tank (length from c. 40cm), with dense clumps of plants to provide shelter and also some floating plants. They should not be kept with large, excitable fishes. The water should not be too deep, because like other labyrinth fishes they have to come to the surface to breathe air. Water temperature about 25°C.—**Diet:** live and dried food, with some greenstuff.—**Breeding:** quite easy.

Kissing Gourami, *Helostoma temmincki*

Anabantidae

Characteristics: up to 30cm, but usually only half this length in the aquarium. The sexes are very difficult to distinguish, although with age the females become somewhat stouter. There are various colour varieties, of which the pink is the one most commonly available on the market. The popular name has no sexual connotation. The so-called kissing is, in fact, a form of sham fighting behaviour. Apart from these harmless confrontations these are very quiet fishes which mostly live in the middle water layers.—**Distribution:** south-east Asia (Thailand, Malaya, Sumatra, Java, Borneo).—**Maintenance:** in a well-planted community tank with open water for swimming. Naturally at least two individuals must be kept if the "kissing" behaviour is to be seen. Water temperature at least 25°C.—**Diet:** on account of the peculiar structure of the mouth they require fine live and dried food, and they must have some vegetable matter.—**Breeding:** very difficult.

Croaking Gourami, *Trichopsis vittatus*

Anabantidae

Characteristics: up to 7cm. These are slender labyrinth fishes which produce croaking sounds, particularly during the courtship period. They are peaceful and rather shy and are normally seen in the middle water layers. The colours are more intense in the males, and the dorsal and anal fins more pointed.—**Distribution:** south-east Asia (Thailand, Vietnam, Malaya, Sumatra, Java, Borneo).—**Maintenance:** in a community tank with dense vegetation to provide hiding-places and open water in the centre. Some floating plants will also be welcome. They should not be kept with large boisterous species.—**Diet:** can be mainly dried food, with some live food.—**Breeding:** rather difficult, and only successful under favourable conditions.

Striped Anostomus, *Anostomus anostomus*

Anostomidae

Characteristics: up to 18cm, but usually less in the aquarium. This is a member of the headstander family, so called because the fish move around for most of the time, and especially when feeding, with the head downwards. The sexes cannot be distinguished externally. These are rather active fishes living in the middle and lower water layers and for most of the time they are very peaceful. However, older individuals may become aggressive when defending a territory, particularly towards other members of their own species.—**Distribution:** South America (Amazon basin and Guyana).—**Maintenance:** in a large community tank with scattered clumps of plants, roots to provide hiding-places and the darkest possible substrate. If the tank is not very spacious only one specimen should be kept, otherwise there will be quarrels. Water temperature about 25°C or slightly higher.—**Diet:** live and dried food with some greenstuff, such as lettuce. They also like to browse algae.—**Breeding:** very difficult.

Anostomus ternetzi

Anostomidae

Characteristics: up to 8cm. This is a rarely imported headstander which is smaller and more peaceful than the preceding species. Otherwise it shows the same type of behaviour. The sexes are very difficult to distinguish, although at certain times the females have a more rounded body.—**Distribution:** South America (Orinoco and Guyana).—**Maintenance:** in a community tank with scattered clumps of plants which should provide sufficient shelter. The composition of the water is evidently not critical. Water temperature about 25°C.—**Diet:** all kinds of live and dried food. As the mouth is dorsal the fish literally lean over backwards when browsing algae from the plants. They also take food from the bottom.—**Breeding:** evidently this has not yet been achieved in the aquarium.

Headstander, *Abramites microcephalus*

Anostomidae

Characteristics: up to 13cm, but less in the aquarium. A very variable species with irregular dark transverse bands. Females have duller coloration than males, and become relatively taller with age. The popular name refers, of course, to the habit of swimming in an oblique position with the head down. In general, headstanders are peaceful, at least when young, but they will attack any delicate plants and occasionally also the fins of the other occupants of the tank.—**Distribution:** South America (Amazon basin).—**Maintenance:** in a large community tank with some dense vegetation and open water for swimming. Plants with soft leaves are not, of course, suitable. Headstanders must not be kept with small or slow fishes. They have a tendency to jump out of the water, so the tank must have a good, close-fitting lid. Water temperature about 25°C or slightly higher.—**Diet:** live and dried food in variety, and some plant matter is essential.—**Breeding:** very difficult, and hitherto only achieved on a few occasions.

Three-banded Pencilfish,
Nannostomus trifasciatus

Hemiodontidae

Characteristics: up to 6cm. This elegant, slender and beautifully coloured fish is unfortunately not often available on the market. It is also rather delicate and thus only suitable for the experienced aquarist. There are no external sex differences, except that the females are normally stouter and less brightly coloured. These are sociable and very peaceful fish which live in the upper and middle water layers.—**Distribution:** South America (Guyana, Rio Negro and Amazon).—**Maintenance:** as a small shoal in a community tank, even a small one, with some dense vegetation. Water temperature about 25°C.—**Diet:** preferably small food, particularly brine shrimps, with dried food only as a supplement.—**Breeding:** very difficult.

Dwarf Rainbowfish, *Melanotaenia maccullochi*

Atherinidae

Characteristics: up to 8cm. These are very peaceful, un-demanding fishes. It is typical that this and other members of the family, generally known collectively as silversides or sand smelts, have two dorsal fins. The sexes can be distinguished quite easily. Females are paler and their dorsal and anal fins are not so pointed as those of the males. These are sociable, shoaling fishes living mainly in the middle and lower water layers.—**Distribution:** northern Australia.—**Maintenance:** always as a shoal in a spacious community tank with scattered clumps of plants and plenty of open water for swimming. Water temperature about 24°C, or slightly lower.—**Diet:** all kinds of live and dried food, although the live food can be omitted. They must, however, have some plant food. —**Breeding:** not too difficult.

Madagascar Rainbow, *Bedotia geayi*

Atherinidae

Characteristics: up to 15cm, but generally much less in the aquarium. This attractive fish is quite suitable for the begin-ner as it is very peaceful and undemanding. The colours are more intense in the males which also have dark red edges to the fins. These are lively shoaling fishes which live in the middle water layers.—**Distribution:** Madagascar.—**Main-tenance:** as a small shoal in a large, preferably long com-munity tank which will give them sufficient open water for swimming. The composition of the water is not critical. Water temperature about 24°C.—**Diet:** omnivorous and they can be fed on dried food only. Some plant matter is, however, ap-preciated.—**Breeding:** not difficult.

Coolie Loach, *Acanthophthalmus kuhlii*

Cobitidae

Characteristics: up to 10cm. There are several forms of coolie loach, all differing somewhat in pattern, but all have the typical eel-like movements. These are attractive fishes for the home aquarium being peaceful and hardy. They live most of the time on the bottom and are often hidden by day.—**Distribution:** south-east Asia.—**Maintenance:** in a community tank with a certain amount of vegetation, but with plenty of shelter on the bottom. At least part of the substrate should consist of soft sand. It is advisable to keep several individuals together. They have a tendency to jump out of the tank when alarmed. Water temperature about 25°C.—**Diet:** live and dried food.—**Breeding:** has probably been achieved on a few occasions.

Clown Loach, *Botia macracantha*

Cobitidae

Characteristics: up to 30cm, but usually considerably less in captivity (8-12cm). An extremely attractive fish, but one that tends to be rather delicate. There are no external sex differences. These are active shoaling fishes which live mostly on or near the bottom, but swim into the upper waters from time to time, often together with the similarly patterned Sumatra barbs. They are peaceful towards other species but somewhat intolerant of members of their own species.—**Distribution:** south-east Asia (Sumatra and Borneo).—**Maintenance:** as a small shoal in a spacious community tank with subdued lighting, and plenty of small caves in the rockwork to provide shelter, together with roots or old coconut shells. At least part of the substrate should be soft sand. Water temperature about 25°C.—**Diet:** live and dried food, algae.—**Breeding:** possibly achieved on a few occasions.

Indian Glassfish, *Chanda ranga*

Centropomidae

Formerly classified under the genus *Ambassis.*— **Characteristics:** up to 7cm, but less in captivity. The main feature is the translucence of the body which appears iridescent greenish or golden, depending upon the angle of the light. The females can be distinguished by their much duller coloration. These are peaceful, somewhat shy fishes, but unfortunately rather delicate and sensitive to any form of disturbance.—**Distribution:** south-east Asia (India, Burma, Thailand).—**Maintenance:** as a small shoal in a well-established community tank, but only with other peaceful species. The substrate should preferably be dark, and planted to provide hiding-places. Water temperature about 24°C.—**Diet:** all kinds of small live food, with dried food only as a supplement.—**Breeding:** rather difficult.

Chanda wolffi

Centropomidae

Characteristics: up to 20cm, but usually only half this length in the aquarium. Similar in general appearance and coloration to the preceding species, but with an elongated second ray in the anal fin. There are no external sex differences. These are rather shy fishes which are easily frightened, but properly cared for they will do well in the aquarium.—**Distribution:** south-east Asia (Thailand, Sumatra and Borneo).— **Maintenance:** as a small shoal in a community tank with other peaceful species and numerous hiding-places. The substrate should be dark. Water temperature about 24°C or slightly lower.—**Diet:** live food in variety is essential. Dried food can be given as a supplement.—**Breeding:** very difficult, and apparently not yet achieved.

Common Hatchetfish, *Gasteropelecus sternicla*

Gasteropelecidae

Characteristics: up to 7cm, but usually less. The popular name refers to the flattened, hatchet-like shape of the body, caused by the great development of the breast region. The sexes are difficult to distinguish, but the females are usually rather larger than the males. These are typical surface-living fishes which are quite hardy once they have become acclimatized.—**Distribution:** South America (Guyana, Amazon, Peru).—**Maintenance:** in a spacious community tank with sufficient open water near the surface and some floating plants. They should be kept with species which occupy the middle and lower water layers. The tank must be well covered to prevent them gliding out, using the pectoral fins as 'wings'. Water temperature about 25°C or slightly higher. —**Diet:** live and dried food.—**Breeding:** not known to have been bred, or if so only on a few occasions.

Glass Catfish, *Kryptopterus bicirrhis*

Siluridae

Characteristics: up to 10cm. An elongated, bilaterally compressed and almost transparent fish which swims in the middle or lower water layers, usually living in small shoals. These are very peaceful fishes, but unfortunately they are sometimes rather delicate. The oblique position in the water is quite normal.—**Distribution:** south-east Asia (Malaya, Thailand, Sumatra, Java, Borneo).—**Maintenance:** as a small shoal in a community tank with scattered clumps of plants and open water for swimming. They should not be kept with restless species. Water temperature about 24°C.—**Diet:** live food in variety, with some dried food as a supplement.—**Breeding:** evidently not yet achieved.

Synodontis flavitaeniatus

Mochocidae

Characteristics: up to 15cm, but less in the aquarium. This is one of the most colourful of the African catfishes and is related to the better known upside-down catfish *Synodontis nigriventris*. It is a peaceful and sociable fish mainly active at twilight and during the night, but also to be seen by day using the long barbels to search for food on the bottom.—**Distribution:** central Africa (Congo region).—**Maintenance:** in a spacious community tank which must be furnished with rockwork and roots to provide shelter, and also a slanting branch on which the fish will rest. The lighting should be subdued and part of the substrate should be soft. Water temperature about 25°C.—**Diet:** live and dried food.—**Breeding:** evidently not yet achieved.

Silver Needlefish, *Xenetodon cancila*

Belonidae

Characteristics: up to 30cm, but usually less. This rather unusual fish is not often available on the market and is really only suitable for the experienced aquarist. It is a peaceful, surface-living species which often remains motionless in the water, but can become very active, even jumping out of the water. The sexes are very difficult to distinguish, but males usually have dark edges to the fins.—**Distribution:** south-east Asia (India, Sri Lanka, Burma, Thailand, Malaya).—**Maintenance:** is possible in a spacious, well-planted community tank with fishes of the same size but which live in the lower water layers. It is preferable, however, to keep several individuals in a large and long species tank which must have a close-fitting lid. Water temperature about 25°C.—**Diet:** mainly live food, including small fishes and frogs.—**Breeding:** evidently not yet achieved.

Figure-eight Pufferfish,
Tetraodon palembangensis
Tetraodontidae

Characteristics: up to 20cm, but normally less in captivity. Members of this family are able to inflate themselves with air or water. They do this when fighting or if alarmed. There are no external sex differences. They normally live in the middle and lower waters and are not really aggressive, although they tend to bite the fins of other fishes with their parrot-like beak.—**Distribution:** south-east Asia (Thailand, Malaya, Sumatra, Borneo).—**Maintenance:** in a large community tank with large, robust fishes, but preferably in a large species tank with tough plants and hiding-places constructed from rocks and roots. Water temperature about 24°C.—**Diet:** mainly live food, particularly earthworms and small snails, with some dried food and vegetable matter.—**Breeding:** very difficult, and probably rarely achieved.

Somphong Pufferfish, *Tetraodon somphongsi*
Tetraodontidae

Sometimes known as *Carinotetraodon somphongsi.*— **Characteristics:** c. 8cm. A rather variable species with the typical shape and habits of the pufferfishes. Sometimes quite peaceful, but at other times it is apt to bite other fishes. During courtship or when excited the back and belly each develop a comb-like appearance. The sexes are difficult to distinguish, although the females generally have duller colours.—**Distribution:** south-east Asia (Thailand).— **Maintenance:** in a community or species tank with plants for shelter and open water for swimming. Water temperature about 24°C.—**Diet:** preferably live food, including small earthworms and snails. Dried food is scarcely taken.—**Breeding:** very difficult.

Goldfish, *Carassius auratus*

Cyprinidae

This and the following few pages deal with a selection of cold-water fishes that can be kept in the aquarium, and of these the goldfish is, of course, the best known.—**Characteristics:** up to 25cm, but normally less in the aquarium. The goldfish is a domesticated form which has existed in China for over a thousand years. Its origins are lost in antiquity, but it is probably derived from one or more mutations of *Carassius auratus* which in the wild is usually olive-green. Goldfish were introduced into Europe in the 17th century, and have now spread all over the world. In addition to the normal slender form breeders have, over a long period of years, raised several other forms, some of which are quite grotesque and indeed scarcely viable. Among those which can be kept successfully in a home aquarium are the veiltails, of which two varieties are shown on the opposite page. These are somewhat more delicate than the ordinary goldfish and are only 10-15cm long.—**Distribution:** originally China westwards to eastern Europe.—**Maintenance:**as a group of several individuals in a large, tall, sparsely planted and well-aerated tank. The substrate should be coarse gravel with a few large, flat rocks. Goldfish can also be kept with other large cold-water fishes. There is no need for a heater if the tank is kept in a living room. The ordinary goldfish can be kept at temperatures in the range 4-20°C (or even slightly higher). Veiltails and the other selected forms require a temperature of not less than 15°C. Ordinary goldfish can also be kept in an outside pool.—**Diet:** all kinds of live and dried food, with greenstuff such as lettuce and possibly algae. Dealers sell special brands of goldfish food.—**Breeding:** not difficult, but producing good offspring requires considerable experience.

Ide or Orfe, *Leuciscus idus*

Cyprinidae

Characteristics: up to 75cm, but less in the aquarium. The females are larger than the males. There has long been a selected or domesticated golden orfe, which is the form most commonly kept in the aquarium; it does not grow quite so long. In general, these are peaceful and hardy shoaling fishes which are very suitable for the beginner to keep in a cold-water tank.—**Distribution:** central Europe, introduced into Britain.—**Maintenance:** in a spacious, unheated community tank, with tough cold-water plants, sand, coarse gravel, a few smooth rocks and some roots. Several individuals can be kept together with other species. The water should be clear, with a temperature not exceeding 20°C.—**Diet:** live and dried food with some greenstuff.—**Breeding:** difficult in a home aquarium.

Bitterling, *Rhodeus sericeus amarus*

Cyprinidae

Characteristics: up to 9cm. The females are paler and they have a long ovipositor. Bitterlings are interesting not on account of their beauty and peaceful nature, but because of their unusual breeding habits. They mostly live in the lower water layers.—**Distribution:** Europe, introduced into Britain.—**Maintenance:** as a pair in a small cold-water tank with tough plants, a sandy substrate and a few rocks. There must also be at least one living swan mussel in which the female lays her eggs with the help of the ovipostor. The male then fertilizes the eggs with a cloud of sperm. The eggs develop within the swan mussel. Bitterlings can also be kept with other cold-water fishes. Water temperature up to 22°C.—**Diet:** live and dried food.—**Breeding:** not difficult.

Roach, *Rutilus rutilus*

Cyprinidae

Not to be confused with the rudd, *Scardinius erythrophthalmus*, which can also be kept in a cold-water aquarium.—**Characteristics:** up to 20cm, sometimes even more. The sexes are difficult to distinguish. These are sociable and hardy fishes which are at first rather shy and liable to panic.—**Distribution:** Europe.—**Maintenance:** in a spacious community tank with robust plants. They do better when kept as a small shoal, and will then lose their initial shyness more quickly. Water temperature up to 20°C. Roach can also be kept in an outside pond.—**Diet:** plenty of dried and live food, with some plant matter.—**Breeding:** difficult and best attempted in a garden pond.

Bleak, *Alburnus alburnus*

Cyprinidae

Characteristics: up to 15cm. This species occurs in slow-flowing and standing waters, often in very large shoals. They can often be seen close to the banks when taking insects from the surface of the water. It is scarcely possible to distinguish the sexes. Bleak are very active, peaceful fishes normally swimming in the upper water layers.—**Distribution:** Europe.—**Maintenance:** a really spacious tank, not too densely planted, but with a large area of open water for swimming. It is advisable to obtain a number of young individuals, which will soon form a shoal. They can be kept with other cold-water species. The tank should not be overstocked otherwise the bleak may suffer from oxygen deficiency. The water should always be aerated. Water temperature not above 18°C.—**Diet:** mainly live food which is taken from the surface. Dried food and plant matter can be given as a supplement.—**Breeding:** very difficult.

Three-spined Stickleback,

Gasterosteus aculeatus

Gasterosteidae

Characteristics: up to 15cm. This is one of the most interesting European fishes. During the breeding period in spring and summer the male shows brilliant coloration, while the somewhat larger female is a dull olive-green. A pair will establish and defend a territory, in which the male builds a nest and later guards the eggs and young. Sticklebacks, including the related ten-spined stickleback *G. pungitius*, are always active and the males become very aggressive when breeding.—**Distribution:** Europe, northern Asia and northern America.—**Maintenance:** as a pair in a well-planted tank. Several pairs can be kept in a really large tank with dense clumps of plants so that the territories are not in sight of one another, but other species should be excluded. Water temperature up to 22°C, but preferably lower.—**Diet:** all kinds of live food, but dried food will usually be ignored.—**Breeding:** not difficult.

Paradisefish, *Macropodus opercularis*

Anabantidae

Characteristics: up to 12cm, but usually less. The males show more intense coloration and more elongated fins than the females. This is one of the few exotic aquarium fishes which can be kept in an unheated tank. It was, in fact, the first tropical species kept in captivity in Europe. It is a hardy and undemanding fish, its only drawback being a tendency to become increasingly aggressive with age.—**Distribution:** south-east Asia (China, Korea, Taiwan, Vietnam).—**Maintenance:** in a spacious and densely planted tank with other tropical species that prefer fairly low temperatures. However, it is best kept in an unheated species tank at approximately room temperature (15-24°C).—**Diet:** live and dried food.—**Breeding:** not difficult.

Aquarium Fish Index

English Names

Latin names